PARTIALITY, TRUTH AND PERSISTENCE

CSLI
Lecture Notes
Number 15

PARTIALITY, TRUTH AND PERSISTENCE

Tore Langholm

CSLI was founded early in 1983 by researchers from Stanford University, SRI International, and Xerox PARC to further research and development of integrated theories of language, information, and computation. CSLI headquarters and the publication offices are located at the Stanford site.

CSLI/SRI International
333 Ravenswood Avenue
Menlo Park, CA 94025

CSLI/Stanford
Ventura Hall
Stanford, CA 94305

CSLI/Xerox PARC
3333 Coyote Hill Road
Palo Alto, CA 94304

Printed in the United States

94 93 92 91 90 89 88 5 4 3 2 1

Library of Congress Cataloging-in-Publication Data

Langholm, Tore, 1957–
Partiality, truth, and persistence / Tore Langholm.
p. cm. – – (CSLI lecture notes ; no. 15)
Revision of the author's thesis – – Stanford University, 1987.
Bibliography: p.

1. Formal languages. 2. Model theory. I. Stanford University. Center for the Study of Language and Information. II. Title. III. Series.
QA267.3.L36 1988
511.3—dc 19 88–15015
CIP

To my parents

Contents

Preface

In the spring and summer of 1984 I participated with Jens Erik Fenstad, Per-Kristian Halvorsen and Johan van Benthem, in a project to develop a framework for studies in linguistic semantics. The research was conducted at Xerox PARC and at the Center for the Study of Language and Information at Stanford University. A revised and expanded version of the outcome was published last year in (Fenstad et al. 1987). Included in this work was the study of a partial model theory, inspired by the situation semantics of Jon Barwise and John Perry.

As a graduate student at Stanford, I continued these studies under the supervision of Jon Barwise. It was a great privilege to work with him, because of the extent as well as the quality of his help. I am very grateful to him for his generosity throughout these years.

The present book is based on my 1987 Stanford dissertation of the same title, but was substantially reorganized and expanded at the University of Oslo in the summer and fall of 1987, while I was supported by a grant from the Norwegian Research Council for Science and the Humanities. During this period I made a short but very fruitful visit to Johan van Benthem and his colleagues at the Institute for Language, Logic and Information at the University of Amsterdam.

A number of people have read preliminary drafts of the book, and made valuable suggestions. Some of these are mentioned above, others include John Etchemendy, Solomon Feferman and David Israel.

All in all, I have benefited immeasurably from the association with a number of people who have been very generous with their help and guidance. Without their cooperation, this book would not have been realized.

Oslo, February 1988

Introduction

The mathematical discipline of model theory deals with the expressive powers of a formal language. How much can be said with such and such linguistic resources? To give precise meaning to questions of this sort, mathematical *models* are introduced. These represent possible constellations of facts, and are taken to *satisfy* some sentences and *falsify* others. Hence to a given sentence corresponds the class of those models that satisfy it. *Expressive powers* can be measured as the ability to represent such classes.

Standard model theory has concentrated on *complete models*, corresponding to complete constellations of facts, the limiting cases where any addition would introduce inconsistencies. A certain tradition in the philosophy of language has been cited as one influence leading to the exclusive emphasis on totality: The meaning of a sentence is equated with its truth conditions, and the truth conditions are taken to correspond with states of the whole world in which the sentence is true.

Perhaps an equally important factor is the common use of model theory to analyze various forms of logical consequence: One sentence is a logical consequence of another if the truth of the latter is a sufficient condition for the truth of the former. The analysis takes the form of a search for counterexamples: Is there a possible scenario in which one sentence is true and the other false? If such a "scenario" exists, it will be embedded in a larger, "total scenario," a possible state of the whole world. (Or so it is assumed.) Now the latter will also serve as a counterexample to the proposed logical consequence. Hence the "total scenarios" alone are really sufficient for this type of analysis, and there is no need to enter the intricacies of partiality.

Various alternative frameworks have been presented where partiality is introduced in one way or another. The motivations are diverse, but in many cases the point of departure is some particular linguistic phenomenon that cannot be captured adequately by the existing framework. Examples are vague predicates, presupposition failure, sortal incorrectness, or the possibility of self-reference and the associated semantical paradoxes. The corresponding types of partiality come with different interpretations, and one should be careful not to confuse them. Indeed, each variety is worthy of a study in its own right.

In this book we shall not be concerned with any of the phenomena listed above; the present studies are not motivated by any linguistic phenomena beyond those recognized by standard model theory. In fact we shall remain very conservative and stick to a simple first order language. But there will be a shift of emphasis: The relation of satisfaction will be deemed worthy of a study in its own right. Since a given sentence will typically describe only a certain *aspect* of a wider subject matter, we should reflect this in our definition of a satisfaction relation. We should look for a satisfaction relation where the matching parties are the sentence and a *smaller unit* of the total model, rather than the whole model itself. For a simple example, suppose we are studying the natural numbers and are faced with a sentence stating that R is right directed: $\forall x \exists y R(x, y)$. Suppose we also know that every natural number has this relation to its successor. Otherwise the behavior of the relation is unknown. In such a case we would be inclined to say, I think, that the sentence is true *by virtue* of the facts at hand: The sentence is made true by a certain *aspect* of the behavior of the relation, independently of any embedding complete state of affairs. Following up this intuition, we introduce *partial* models where some possible facts are represented, others are not. In this way we can isolate the aspect of a total model that is most directly responsible for the truth or falsity of a given sentence.

These ideas are guided by the ideas behind *situation semantics*: A sentence is true by virtue of a particular situation, a *part* of the world. Situation semantics is a far more ambitious and radical enterprise than the present modest program, with analyses for a number of complex natural language phenomena. It is far beyond the present scope to deal with any but the very simplest of these. We shall concentrate on a simple first order framework, made partial. A number of simplistic assumptions are built into a framework like this, apparently contradicting the complexity recognized by situation semantics. However, these assumptions represent idealizations rather than alleged facts about language phenomena. A number of the idealizing assumptions will be mentioned along the way. These assumptions enable us to concentrate on one particular issue, and to pursue it to an extent that would not otherwise be possible in such limited space.

If the partial models are taken to represent the aspects of a total model that are most directly responsible for the truth or falsity of various sentences, then some partial models will correspond to aspects that are *not* responsible for the truth or falsity of a given sentence. Hence the *structural* partiality gives rise to a *semantical* partiality: Some sentences are neither true nor false relative to a given partial model. This means that we shall be concerned with a type of semantical partiality that is associated with absence of *facts* or *information about facts.* Two interpretations present themselves: A situation, a part of the world, *contains* various constellations of facts, and thereby *makes* some sentences true, some false, while for other sentences again it may not contain a sufficient constellation of facts. Similarly, a given body of knowledge will entitle us to assert some sentences, reject others, while the data could be insufficient to decide on others again. Both interpretations will be considered; the partial models may represent parts of the world or partial information about the world. We shall consider various alternative truth definitions motivated by the two interpretations.

Determinability

Partial model theory is a *generalization* of standard model theory, and provides a richer framework. In particular, it provides a framework where the connection between structural and semantical partiality can be given a first analysis. Principles can be formulated that may or may not govern such a connection; the study of such principles and their interaction with more familiar logical principles will form the subject matter of this book. On the next few pages are presented some plausible principles of partial model theory. Regardless of whether they are accepted as universally valid, it should be clear that they deserve a mathematical scrutiny; in particular it is of interest to identify the limits that they impose.

Determinability is one of these principles. It says that if a sentence is neither true nor false in a model, then this can always be traced back to a structural partiality: 'Sentence undetermined' means 'insufficient facts' or 'insufficient data,' depending on the interpretation. Such an assumption would not fit a framework used to analyze presupposition failure or sortal incorrectness. When such phenomena are allowed for, a sentence could lack a truth value because there is something inappropriate about the use of the sentence in a particular context. The sentence fails to express a definite claim to be verified or falsified, and no additional facts or information will be of any help. It is possible to build frameworks incorporating both types of indeterminateness, but in such an enterprise I think one should be careful not to *conflate* the two types; it should always be clear whether something is nonsensical or just unsupported and undenied by facts. The two varieties of indeterminateness will manifest themselves at different points in the complex interrelationship involving the utterance of a sentence and the evaluation of a definite claim. To equate the two forms would be to give up information that could prove to be important for the analysis of the semantical process. Whereas the gaps introduced by insufficient facts/data will close when the appropriate facts/data are supplied, this is not the case with gaps introduced by presupposition failure or sortal incorrectness. We only intend to analyze the first variety of indeterminateness, and shall be content to design a framework where other gap producing phenomena like presupposition failure are assumed to have already been sifted out. Hence we shall be justified to assume determinability within the framework.

A similar remark applies to vagueness. The existence of vague predicates is a genuine source of indeterminateness, and a very important source at that, but again it is a *different* source. A sentence involving a vague predicate may remain undetermined when all the relevant non-linguistic information is supplied, simply because it is not clear what the predicate *means.* In this case partiality can be traced back to incomplete *truth conditions* rather than incomplete *input to* the truth conditions: It may be unclear to us whether a certain object is blue, for instance, either because we have not seen the object or because the hue so much approaches green that our judgement falters. The distinction is an important one, but also one that is not easily made without adding considerable structure to our simple model theoretic framework. We shall only be concerned with one type, however, and will leave it to a more comprehensive account to compare the two within a single framework.

We shall have nothing to say about semantical paradoxes. Virtually all proposed analyses of the liar paradox introduce partiality in one way or another, but again it is a different type of partiality. In *our* framework all sentences will be *grounded*: They are directly concerned with some feature of the model. When the model is completed, everything is known about these features and the truth or falsity of all sentences are sorted out neatly and independently. Hence not only shall we assume the semantical gaps to disappear *when* sufficient data are supplied, the idea that sufficient data *can* be supplied is assumed to be a coherent one. It may be worth stressing again that the models to be considered here are for almost all purposes ordinary first order ones. Partiality is introduced to give, it is hoped, a clearer picture of the satisfaction relation, but the framework is still sufficiently orthodox to allow the complete models to be included as special cases.

Persistence

Within our model theoretic framework, we shall define a formal language to be *persistent* if the true sentences remain true and the false ones remain false when more facts/data are supplied. It is a trivial matter to design formal languages that violate such a condition. It is less trivial to decide if the occurrences of non-persistence in such languages can be said to reflect corresponding natural language phenomena: Are natural languages persistent? A reasonable answer is *yes*, but the question needs clarification: The interpretation of natural language sentences will normally depend on the context of use; such sentences are rarely true or false in and of themselves. The model theoretic framework represents an idealization where this dependence is eliminated, but in the general case a constellation of facts will satisfy or falsify not a sentence but rather a claim expressed by the sentence in a particular context. Hence persistence should apply to *claims* rather than sentences. So in the general case a language is persistent iff the true claims remain true and the false claims remain false when more facts/data are supplied.

Many phenomena look like instances of non-persistence. For instance, it can certainly happen that a sentence of a natural language, when uttered in a restricted context situation, will express a true claim, while the same sentence will express a false claim when instead it is uttered in a wider context situation. But as long as the two claims are not the same, this would not be an example of non-persistence. For instance, the domain of reference for a universally quantified sentence could be contextually determined, and interpreted differently in the two cases. When the more restricted domain is referred to, the universally quantified sentence would express a weaker claim than it would if instead a wider domain was referred to.

So it can certainly happen that a *sentence* displays a non-persistent behavior relative to situations of context, the situations that help provide the interpretation of the sentence. But once the interpretation is settled, once a definite claim is made, the sentence has a persistent behavior relative to the situation supporting its truth.

There is a lot more to be said about claims, utterances, sentences, propositions, etc. We have not even said what a claim *is*; in the above example we relied on a pretheoretic understanding of the term. When the question of persistence is pushed a little further than it was in the above example, then one will soon discover that the question can only receive a satisfactory treatment *within* a particular theory of language, or at least within a theory that tells us a little more about how exactly one should understand the abstract notion of a claim, or of a proposition.

No such definition will be attempted here, since it does not really concern the studies we are about to undertake. Regardless of which position one chooses to take in the philosophy of language, we will have made our point: Persistence, whether or not a universal condition on languages, is a *significant* condition well worth a mathematical study.

Coherence and Reliability

Finally, in addition to the two conditions of determinability and persistence, we also introduce the obvious condition of *coherence*: A given sentence is never both true and false in the same model.

In a general setting, these conditions of coherence, determinability and persistence define *relations* between sentences and truth definitions. When the language is given, a truth definition is persistent iff every sentence of the language is persistent relative to that truth definition. Conversely, when the truth definition is given, we can divide the sentences of a language into the persistent and the non-persistent ones, and a language is persistent iff all of its sentences are.

We define a fourth property which relates a given truth definition for our specific first order language to the classical truth definition for that language: A truth definition is *reliable* if the truth or falsity of a sentence in a partial model implies the truth or falsity, in the standard sense, of that sentence in all the completions of the model. When persistence and determinability are already assumed, reliability amounts to the obvious condition that the classical truth definition should apply when information is complete. We note that reliability implies coherence. Still, coherence is a useful notion in its own right since unlike reliability it applies to arbitrary languages for which no standard truth definition is already established.

Research Topics

The present research is guided by questions in two different areas. First, with the introduction of gaps in the semantics, there are various ways to generalize the classical truth definition for a simple first order language: We know what it means for a sentence to be true or false in a classical, complete model, but how do we extend this relation when the partial models are introduced? Various alternatives exist, and we shall make a detailed comparison between a number of these.

The second group of questions are of a more conventional model theoretic orientation; we shall gauge the expressive powers of various formal languages, and carry out some comparisons. But also in this area new issues arise: Coherence, determinability and persistence represent limitations on the expressive power of a formal language; they are negative expressibility properties. Now, how do they interact with more familiar expressibility properties studied by classical model theory?

At this point, an example from propositional logic will suffice: The connectives $\vee$ and $\neg$ are, under a certain reasonable truth definition $\models$, not functionally complete with respect to a three valued logic with truth values "true," "false" and "gap." From a theoretical point of view, it is of interest to identify *extensions* of the set $\{\vee, \neg\}$ that are complete in this sense. Such completeness violates persistence, however, so it is *also* of interest to identify complete sets of connectives relative to the set of truth functions that are coherent, determinable and persistent.

The two themes are intertwined, and there is a considerable overlap. By a *language* we shall mean a syntactic component, a *vocabulary*, coupled with a matching truth definition for the sentences of that vocabulary. Now, we shall acquire insights about various alternative truth definitions $\models$, $\models\!\!=$, $\models\!\!=_\omega$, $\models_\Box$, etc., for our simple first order vocabulary $\mathcal{L}$. These alternative truth definitions yield different languages $\langle \mathcal{L}, \models \rangle$, $\langle \mathcal{L}, \models\!\!= \rangle$, $\langle \mathcal{L}, \models\!\!=_\omega \rangle$, $\langle \mathcal{L}, \models_\Box \rangle$, and knowledge about $\models\!\!=$, for instance, will transform into knowledge about the language $\langle \mathcal{L}, \models\!\!= \rangle$.

A second and less obvious example of the connections between the two areas is provided by the relative saturation theorem. This result is obtained through a study of alternative truth definitions, but finds an important application in a result about extended languages.

Since two different interpretations will be considered for the partial models, one should be open for the possibility that different relations of satisfaction between language and model are appropriate, depending on which interpretation is intended. The sentential operator $\vee$ will correspond to the inclusive 'or' of informal language. Hence $(\varphi \vee \psi)$ is true iff φ or ψ is true. When truth relative to situations is taken as a basic semantical relation, then a plausible stipulation would rule that $(\varphi \vee \psi)$ is true in a particular situation iff φ or ψ is true in that situation. This behavior of $\vee$ is the distinguishing feature of the strong Kleene truth definition, which has been embraced by a number of authors. In the analysis of naked infinitive perception reports in Barwise (1981), *scenes* are identified as the objects of perception. A scene is a situation of a special type, and a subject *sees* φ just in case she sees a scene in which φ is true. On this analysis, empirical data on the semantics of various items of language are available in the form of our intuitive reasoning about naked infinitive perception reports. Since to see $(\varphi \vee \psi)$ is to see φ or to see ψ, it is plausible to infer that $(\varphi \vee \psi)$ is true in a situation iff φ or ψ is true in that situation. And this is the crucial feature of the strong Kleene truth definition. As a consequence, a model containing no information about a sentence φ does not make the sentence $(\varphi \vee \neg\varphi)$ true in the strong Kleene sense. And in general, the strong Kleene truth definition is not closed under classical consequence: If φ is true in a model, in the strong Kleene sense, and ψ is a classical consequence of φ, then it does not follow that also ψ is true in this model in the strong Kleene sense.

While the strong Kleene truth definition is a likely candidate for the satisfaction relation under the first interpretation of the framework, things become different when we turn to the alternative interpretation where a partial model corresponds to partial information, and the relation of satisfaction corresponds to assertability on the basis of information obtained. Now *information* does not make anything true; with a piece of information we can deduce that a given sentence must be true, but ultimately the sentence is *made true* by facts rather than information. Hence on this alternative interpretation, the relation of satisfaction between sentences and models is not meant to represent a basic semantical relation. Rather, the truth of a sentence follows from a piece of information iff this piece of information implies the existence of a situation that makes the sentence true. Hence we should be open also for truth definitions that are more indirect in spirit, like the definition obtained from the *supervaluation schema*: One can define truth in a partial model as truth in *all* complete models into which the partial model can grow. This definition was introduced originally by Bas van Fraassen in his work on free logic and on the liar paradox, cf. van Fraassen (1968). Clearly this truth definition is reliable. We also notice that it is maximal in this respect: Let one truth definition be *at least as strong as* another if the truth or falsity of a sentence relative to the latter always implies the truth or falsity, respectively, of that sentence relative to the former. Then the supervaluation truth definition is at least as strong as any reliable truth definition. Hence we shall call it the *strongest reliable truth definition.*

The strong Kleene truth definition and the strongest reliable truth definition do not turn out to be equivalent. They will agree on the introductory example with the sentence expressing that R is right directed, but in general the matter is not so straightforward: The complete models follow a classical logic, hence the supervaluation truth definition, unlike strong Kleene, is closed under classical consequence. And moreover, sometimes when a sentence is true in all the completions of a partial model, the mechanisms involved are highly complex and roundabout, and are not naturally accounted for "locally" in the partial model. In fact, these "highly complex mechanisms" are sometimes of a second order nature: In the supervaluation truth definition there is a quantification over all possible completions of a model, equivalently over all complete relations that the partial relations of the model can grow into. It has already been noted (cf. Woodruff 1984) that a Skolem–Löwenheim result fails for the supervaluation truth definition in a particular framework for free logic. Such properties of course depend on how the framework is set up, but also in the present framework it turns out that the supervaluation truth definition does not satisfy a fairly standard Skolem–Löwenheim result.

One motivation for supervaluations over strong Kleene is the closure under classical consequence, but the violation of a Skolem–Löwenheim property seems like a rather drastic consequence of such a modest consideration. And it turns out that if we look closer, we find several rather nice truth definitions between strong Kleene and supervaluation, that are both closed under first-order consequence *and* satisfy a Skolem–Löwenheim result. As we study these truth definitions, we asses the extent to which strong Kleene and supervaluations differ, and we try to trace the difference to various formal properties of the truth definitions. We shall "experiment" with strong Kleene; the intermediary truth definitions will represent various natural strengthenings of this truth definition.

Formal Framework

We now turn to the specifics of our framework. The partial models to be considered will all specify a definite domain of individuals. Furthermore they will specify, for each relation symbol, a positive and a negative extension. The extensions contain n-tuples of individuals, and correspond to elements that *do* and *do not* relate to each other in the sense of the particular relation symbols. The two extensions are disjoint, but do not have to be exhaustive. About the remaining n-tuples the model does not provide information. Standard models can be viewed as partial models where for each relation symbol the positive and negative extensions together *are* exhaustive. One model will be defined as an *informational* extension of another if their domains of individuals *coincide*, and the positive and negative extensions of the latter are, one for one, contained in their counterparts in the former.

Hence we are restricting our attention to partial aspects that are complete as far as the domain of individuals is concerned. Stated in terms of information content, we are dealing with partial information against a background where the domain of individuals is well known and understood. Hence all our discussions will be concerned with a somewhat special case. We have chosen this notion of a partial model since it seems to represent the minimal modification of its standard counterpart that at all allows for the representation of partial information. It turns out that this modification alone introduces a number of phenomena and distinctions that cannot be represented in classical model theory. Hence it would seem a natural first step to investigate the basic features of such a modestly generalized model theory. When more generalizations are introduced, the findings will represent generalizations of our present results.

A similar remark applies to the fact that we shall consider only similarity types without function symbols or individual constants. By leaving out such symbols we deliberately ignore a number of issues about partiality that certainly have their place in a wider analysis. On the other hand, the questions we shall raise are specifically concerned with the behavior of various truth definitions relative to various logical operators, and it is hoped that the relevant questions and results will be better perceived if we leave out more peripheral and supposedly distracting issues about partiality of constants and functions, how identity interacts with partially defined entities, etc.

Overview

Chapter 1 on propositional logic serves as an additional introduction to the subject. A number of the main issues are addressed in a simplified form, and simplified versions are presented of notions and arguments that are introduced in a more general predicate logic setting later in the book. Thus to a certain degree, chapter 1 is a "mini version" of chapters 2–4, and it is very similar in structure.

Chapter 2 gives an outline of the elementary properties of the strong Kleene truth definition. Most of the results, as well as the methods used, will be very straightforward generalizations of their counterparts in standard model theory. There will be some unexpected twists, however. Notably, the Craig interpolation theorem appears as an almost trivial result, and we shall identify an algebraic characterization of semantical equivalence which does not represent the most straightforward generalization of the classical Ehrenfeucht–Fraïssé criterion. The familiar symbol $\models$ will be reserved for the strong Kleene truth definition.

Many of the results will be obtained through a translation procedure which allows us to reduce certain questions about $\models$ to questions about standard first order logic. This strategy is identical to that found in Feferman (1984), which again was a generalization of the strategy used in Gilmore (1974). In this chapter much of the groundwork is done for the proofs in chapters 3 and 4, where the main issues of the book are addressed.

Chapter 3 contains an extensive comparison between various alternative truth definitions. We start out with the fact that $\models$ is not closed under classical first order consequence, and define $\models^{wscl}$, the *weak syntactic closure* of $\models$, as the closure of $\models$ under classical consequence. The *strong syntactic closure*, $\models^{sscl}$, on the other hand, is the simultaneous closure of $\models$ under classical consequence and universal quantification. $\models^{wscl}$ and $\models^{sscl}$ bear a certain similarity to the strongest reliable, or "supervaluation" truth definition: In propositional logic they all coincide, and the three of them represent alternative generalizations to predicate logic of the supervaluation truth definition for pure propositional logic. At the same time, we identify a modification of $\models$ obtained by forsaking the compositionality features. The notation '$\models\!\!=^*$' is used for this truth definition. The schema by which we define $\models\!\!=^*$ is quite naturally construed as a generalization of the schema by which $\models$ is defined. Hence we shall call it the *generalized strong Kleene* truth definition.

The generalized strong Kleene truth definition is stronger than the strong Kleene truth definition. We prove that $\models\!\!=^*$ coincides with $\models^{sscl}$, and that a natural weakening of $\models\!\!=^*$ coincides with $\models^{wscl}$. Moreover, we prove that $\models\!\!=^*$ satisfies a strong version of the Skolem–Löwenheim theorem, and that the generalized strong Kleene truth definition and the strongest reliable truth definition *coincide* on countable structures. As a corollary to these two results, it follows that the generalized strong Kleene truth definition is the strongest reliable truth definition satisfying a particular version of the Skolem–Löwenheim theorem.

The analogue to a necessity operator in possible worlds semantics suggests the notation '$\models_\Box$' for the supervaluation truth definition. A result by P. Woodruff was mentioned above; he shows that both the upwards and downwards Skolem–Löwenheim theorems fail for the supervaluation semantics for free logic. In our framework things will turn out slightly different from the way they do in free logic, and $\models_\Box$ *does* in fact satisfy a simple version of the (downwards) Skolem–Löwenheim theorem. However, $\models_\Box$ will fail to satisfy a familiar strengthening involving the notion of substructures. It will follow that $\models^*$, although stronger than $\models$ and $\models^{wscl}$, is weaker than $\models_\Box$.

Next we return to a comparison between $\models$ and $\models_\Box$ from a different perspective: We shall see that $\models$ and $\models_\Box$ do not coincide for formulas in general, but certainly they will be equivalent for some formulas. Such formulas we call *predictive.* One might speculate that every formula is classically equivalent to a predictive formula, or even to a formula from a recursive set of predictive formulas. Such a result holds in propositional logic, but in predicate logic it fails. The fact that it *does* fail can be seen from the fact noted above, that $\models$ does not have the expressibility properties of $\models_\Box$. However, we shall consider some weakenings of this notion of a predictive formula, and try to determine how much it has to be weakened in order that every formula is classically equivalent to a formula with such a property. These investigations culminate with the *relative saturation theorem.*

Finally, in chapter 4 we apply some of the results from the preceding chapters to obtain various characterization results for extensions of the language: With the strong Kleene truth definition, all formulas of the basic first order language are coherent, determinable and persistent. Now it is possible to extend the language with new compositional, propositional connectives. In particular, a non-persistent symbol for *exclusion negation* is often introduced since this adds considerably to the expressive power of the language. It turns out, however, that when interpreted by the strong Kleene truth definition, the language *without* exclusion negation is in a certain sense complete with respect to coherence, determinability and persistence. Of course, such completeness will only hold relative to some equivalence relation on sentences. Two sentences are *strongly equivalent* if they are true in the same models, *and* false in the same models. We show that in the extension of the language obtained by adding exclusion negation, every persistent sentence is strongly equivalent to a sentence of the original language. Moreover, in *any* extension of the language obtained by adding compositional, propositional connectives, any coherent, determinable and persistent sentence will be strongly equivalent to a sentence of the original language. Some results of this type have been proved earlier, in Blamey (1980) and Fenstad et al. (1987). There are interesting differences in the various methods of proof; the present strategy using results about alternative truth definitions has not been presented earlier. Finally, we prove a combined Lindström and persistence characterization theorem: The basic first order language is in a certain sense *maximal* with respect to coherence, determinability, persistence and two additional properties familiar from classical first order logic, namely a compactness property and a Skolem–Löwenheim property.

1

Propositional Logic

In this chapter we address some questions about propositional connectives in a partial setting. Hence we restrict our attention to a simplified language without quantifiers. Structurally the chapter is very similar to the remainder of the book, where the same issues are taken up in a more general predicate logic framework. Hence the chapter will serve as a "first run" through a simplified version of the main arguments of the book. But propositional logic is of course more than a toy version of predicate logic. Results in propositional and predicate logic do not always parallel each other, and several proofs are included in this chapter for which there are no direct counterparts in predicate logic.

1.1 Basic Notions

We study a simple language $\mathbf{l}$ for propositional logic, with the 0-ary, 1-ary and 2-ary connectives $\top$, $\neg$, $\vee$. A *similarity type* for this propositional logic is a finite set ρ of sentence symbols S. The set $\mathbf{l}[\rho]$ of $\mathbf{l}$ sentences of similarity type ρ is built up from $\top$ and the sentence symbols of ρ, using $\neg$ and $\vee$ in the usual way.

A *model* v is a triple $\langle \rho_v, v^+, v^- \rangle$, where ρ_v is the similarity type of v, and v^+ and v^- are disjoint subsets of ρ_v. We choose this format rather than the simpler alternative $\langle v^+, v^- \rangle$, since in the latter case a model for ρ would also be a model for any $\rho' \supseteq \rho$. The difference is not an important one, but in some cases precision will be better served by the chosen alternative.

We say that v *informationally extends* u, $u \ll v$, if $\rho_u = \rho_v$, $u^+ \subseteq v^+$ and $u^- \subseteq v^-$.

A model v is *complete* if $v^+ \cup v^- = \rho_v$. If $u \ll v$ and v is complete, then v is an *informational completion* (or just *completion*) of u.

In the sequel, unless otherwise indicated we shall assume that some particular similarity type has been singled out, and only models and sentences of that particular similarity type are being considered.

We shall study several *truth definitions*, several alternative notions of what it means for a sentence to be true or false in a partial model. In his 1938 article *On notation for ordinal numbers*, S. Kleene introduced a truth definition which is now known as the *strong Kleene* truth definition. We reserve the familiar symbol '$\models$' for this truth definition. Truth and falsity relative to $\models$ are defined in parallel; '$v \models \varphi^+$' expresses that the sentence φ is true in v, while '$v \models \varphi^-$' expresses that φ is *false* in v:

$$v \models S^+ \text{ iff } S \in v^+.$$

$$v \models S^- \text{ iff } S \in v^-.$$

$$v \models \top^+.$$

$$v \not\models \top^-.$$

$$v \models \neg\varphi^+ \text{ iff } v \models \varphi^-.$$

$$v \models \neg\varphi^- \text{ iff } v \models \varphi^+.$$

$$v \models (\varphi \vee \psi)^+ \text{ iff } v \models \varphi^+ \text{ or } v \models \psi^+.$$

$$v \models (\varphi \vee \psi)^- \text{ iff } v \models \varphi^- \text{ and } v \models \psi^-.$$

This particular format may have originated with Hans Kamp, cf. van Benthem (1984). Kleene himself used an equivalent truth table presentation; we shall look at this shortly.

Sometimes we shall simplify notations, and write '$v \models \varphi$' for '$v \models \varphi^+$'. Hence when the polarity sign is suppressed, it is assumed to be positive. We write '$\models_2$' for the restriction of $\models$ to complete models. Hence if $v \models_2 \varphi^+$ or $v \models_2 \varphi^-$, then v is complete. Moreover, we shall use '$\models$' and '$\models_3$' interchangeably. The subscripts '2' and '3' refer to the numbers of possible "truth values" in the resulting logics. As we shall prove below, complete models behave like classical ones, and yield a two-valued logic. With the partial models there are three possibilities, truth falsity and *neither*, also referred to as *gap*. We refrain from calling the associated logic "three-valued" since the gap does not quite play the rôle of a truth value. On the other hand, these two numbers of possibilities give more than sufficient mnemonic reason for the notation '$\models_2$' and '$\models_3$'.

Theoretically there is a fourth possibility; at a later point we shall study a generalized framework where sentences can be *both* true and false at the same time. Due to the above stipulation that v^+ and v^- are disjoint, this never happens with the proper models; under the truth definition $\models$ the language **l** is *coherent*:

Theorem. *For any model v and* $\mathbf{l}$ *sentence φ, it is not the case that both $v \models \varphi^+$ and $v \models \varphi^-$.*

Proof: By induction on sentences.

- $v \not\models \top^-$.
- By the *structural* coherence condition on v we do not have both $S \in v^+$ and $S \in v^-$. Hence neither do we have both $v \models S^+$ and $v \models S^-$.
- If $v \models \neg\varphi^-$ and $v \models \neg\varphi^+$, then both $v \models \varphi^+$ and $v \models \varphi^-$, so we have a conflict with the induction hypothesis.
- If $v \models (\varphi \vee \psi)^-$ and $v \models (\varphi \vee \psi)^+$, then $v \models \varphi^-$, $v \models \psi^-$, and either $v \models \varphi^+$ or $v \models \psi^+$. In either case, there is a conflict with the induction hypothesis. □

Thus we have justified the suffix in '$\models_3$'. The suffix in '$\models_2$' is justified by a dual result, which expresses *determinability*:

Theorem. *If v is a complete model, then for any* $\mathbf{l}$ *sentence φ either $v \models \varphi^+$ or $v \models \varphi^-$.*

The proof closely parallels the proof of the previous lemma, and is left as an exercise. Finally we note that *persistence* holds. From here on, we use '$\odot$' as a varible to range over $+$ and $-$.

Theorem. *If $u \ll v$ and φ is an* $\mathbf{l}$ *sentence, then $u \models \varphi^\odot$ implies $v \models \varphi^\odot$.*

Proof: By induction on sentences.

- This is trivial for $\top$. For sentence symbols S it follows immediately from the definition of $\ll$, together with the appropriate clause in the definition of $\models$.
- Suppose $u \models \neg\varphi^+$. Then $u \models \varphi^-$ and hence $v \models \varphi^-$ by the induction hypothesis. Consequently $v \models \neg\varphi^+$.
- Suppose $u \models \neg\varphi^-$. Then $u \models \varphi^+$ and hence $v \models \varphi^+$ by the induction hypothesis. Consequently $v \models \neg\varphi^-$.
- Suppose $u \models (\varphi \vee \psi)^+$. Then either $u \models \varphi^+$ or $u \models \psi^+$. In either case we get $v \models \varphi^+$ or $v \models \psi^+$ by the induction hypothesis. Hence $v \models (\varphi \vee \psi)^+$.
- Suppose $u \models (\varphi \vee \psi)^-$. Then both $u \models \varphi^-$ and $u \models \psi^-$. By the induction hypothesis both $v \models \varphi^-$ and $v \models \psi^-$, and hence $v \models (\varphi \vee \psi)^-$. □

The strong Kleene truth definition for $\mathbf{l}$ gives rise to a corresponding consequence relation. Let Γ be a set of signed sentences and let ϕ be a signed sentence. We define $\Gamma \models_3 \phi$ to hold iff $v \models_3 \phi$ for all models v such that $v \models_3 \phi_0$ for all $\phi_0 \in \Gamma$.

We have assumed a fixed similarity type ρ; hence in '$\Gamma \models_3 \phi$' we are quantifying over ρ models. With more than one similarity type around, ambiguities could arise. Suppose ρ' is an extension of ρ. When ρ and ρ' are studied side by side, a disambiguating index for ρ or ρ' would seem to be needed. Thus in '$\Gamma {\models_3}^{\rho} \phi$' and '$\Gamma {\models_3}^{\rho'} \phi$' we are quantifying over ρ models and ρ' models respectively. But the two are equivalent. This is seen to follow from the next lemma, which involves the notion of a *restriction* of a model to a smaller similarity type. Suppose $\rho \subseteq \rho'$ and v is a ρ' model. Then $v \uparrow \rho$ is the triple $\langle \rho, v^+ \cap \rho, v^- \cap \rho \rangle$. The lemma follows by a simple induction on sentences, and is left to the reader. We follow up the practice, initiated a few paragraphs above, of using 'ϕ', 'ϕ'', etc., to range over signed sentences. Moreover, we take the liberty of writing '$\phi \in \mathbf{l}[\rho]$' rather than the more accurate '$\phi \in \{\varphi^+ \mid \varphi \in \mathbf{l}[\rho]\} \cup \{\varphi^- \mid \varphi \in \mathbf{l}[\rho]\}$'.

Lemma. *If v is a ρ' model, $\rho \subseteq \rho'$ and $\phi \in \mathbf{l}[\rho]$, then*

$$v \models_3 \phi \quad \textit{iff} \quad v \uparrow \rho \models_3 \phi.$$

Analogously to '$\Gamma \models_3 \phi$' we also write '$\Gamma \models_2 \phi$'. In the latter we restrict to complete models, so in the special case when ϕ and all members of Γ are positively signed, this expresses classical consequence. A simple lemma is called for to justify this assertion. Let 'v_{cl}' range over classical models for a given similarity type ρ. v_{cl} is a subset of ρ, alternatively represented as the characteristic function for this subset relative to ρ. Consider the relation $\bowtie$ between classical models v_{cl} and complete partial models v:

$$v_{cl} \bowtie v \quad \text{iff} \quad v_{cl} = v^+$$

Since the relation is only defined for complete models v of the partial format, and *they* all satisfy $v^- = \rho - v^+$, this is a 1-1 relation. Let $\models_{cl}$ be the standard satisfaction relation for classical models for propositional logic:

Lemma. *If $v_{cl} \bowtie v$, then for every $\mathbf{l}$ sentence φ we have*

$$v_{cl} \models_{cl} \varphi \quad \textit{iff} \quad v \models_2 \varphi.$$

Proof: Since v is complete, we know by the above coherence and determinability results that $v \models_2 \varphi^-$ iff $v \not\models_2 \varphi^+$, and hence that $v \models_2 \neg\varphi^+$ iff $v \not\models_2 \varphi^+$. Replacing the given positive clause for negation in the definition of $\models$ with *this* clause, we see that the definition of $\models_{cl}$ *coincides* with the definition of $\models$, as restricted to positively signed sentences φ^+. Hence the lemma follows. □

In view of this correspondence we shall in the sequel talk about the complete partial models *as if* they were classical models, and take advantage of results from classical propositional logic whenever the opportunity arises.

1.2 Compositionality

Although the connectives $\top$, $\neg$, $\vee$ are the only *primitive* such symbols of **l**, it will be convenient to introduce a few extra, non-primitive connectives. We use '$\bot$' and '$(\varphi \wedge \psi)$' as abbreviations for '$\neg\top$' and '$\neg(\neg\varphi \vee \neg\psi)$' respectively. Also, we could add a primitive symbol '$\sim$' for *exclusion negation*, with semantical clauses

$$v \models {\sim}\varphi^+ \quad \text{iff} \quad v \not\models \varphi^+$$

$$v \models {\sim}\varphi^- \quad \text{iff} \quad v \models \varphi^+$$

Note that the semantic conditions for $\sim$ and $\neg$ coincide on complete models, so $\sim$ and $\neg$ represent alternative generalizations of classical negation. Also observe that the addition of $\sim$ to $\{\top, \neg, \vee\}$ is equivalent to the addition of a unary connective A for *external assertion*, such that $A(\varphi)$ is true iff φ is true, and false otherwise. For clearly $A(\varphi)$ can be defined as ${\sim}{\sim}\varphi$, and ${\sim}\varphi$ as $\neg A(\varphi)$. Exclusion negation is also known as *external negation*. For some additional remarks on these "external connectives," see Bochvar (1981) and Urquhart (1986).

With this extra resource we can define $(\varphi \supset \psi)$, $(\varphi \equiv \psi)$ and $(\varphi \rightleftharpoons \psi)$ as abbreviations for $({\sim}\varphi \vee \psi)$, $(\varphi \supset \psi) \wedge (\psi \supset \varphi)$ and $(\varphi \equiv \psi) \wedge (\neg\varphi \equiv \neg\psi)$ respectively. Note that $\models_3 \varphi \supset \psi$ holds if and only if $\varphi \models_3 \psi$. Hence $\models_3 \varphi \equiv \psi$ iff φ and ψ are true in exactly the same models. In this case we say that φ and ψ are *positively equivalent*. Analogously, φ and ψ are *negatively* equivalent iff $\models_3 \neg\varphi \equiv \neg\psi$, i.e., iff the two sentences are false in exactly the same models. If $\models_3 \varphi \rightleftharpoons \psi$, then we say that φ and ψ are *strongly* equivalent. Hence two sentences are strongly equivalent iff they are both positively and negatively equivalent.

'$(\varphi \equiv \psi)$' and '$(\varphi \rightleftharpoons \psi)$', etc., are useful pieces of notation which we shall use even when φ and ψ themselves are assumed to contain only the connectives $\top$, $\neg$, $\vee$. When C is a set of connectives, we use '$\mathbf{l}_C$' to denote the language where the elements of C have been added to $\top$, $\neg$, $\vee$ as primitive connectives. We allow the notation '$\mathbf{l}_{c_1,\ldots,c_n}$' as an abbreviation of '$\mathbf{l}_{\{c_1,\ldots,c_n\}}$'. Despite the occasional reference to such larger languages, however, 'sentence' will still mean **l** sentence, and 'φ', 'ψ', 'χ' will range over such sentences when not otherwise indicated.

In order to talk about *compositionality* it is convenient to introduce the "truth values" 1, 0 and $\wr$ of sentences relative to models:

$$
\begin{array}{llll}
[\![\varphi]\!]_v = 1 & \text{if} & v \models \varphi^+ & (\text{and } v \not\models \varphi^-) \\
[\![\varphi]\!]_v = 0 & \text{if} & v \models \varphi^- & (\text{and } v \not\models \varphi^+) \\
[\![\varphi]\!]_v = \wr & \text{if} & v \not\models \varphi^+ & \text{and } v \not\models \varphi^-
\end{array}
$$

Using these definitions, we can express the semantical clauses for $\neg$ and $\vee$ with *truth tables*:

φ	$\neg\varphi$
1	0
0	1
$\wr$	$\wr$

φ	ψ	$(\varphi \vee \psi)$
1	1	1
1	0	1
1	$\wr$	1
0	1	1
0	0	0
0	$\wr$	$\wr$
$\wr$	1	1
$\wr$	0	$\wr$
$\wr$	$\wr$	$\wr$

We recall that φ is coherent if $v \models \varphi^+$ and $v \models \varphi^-$ are mutually exclusive. An n-ary connective is a symbol c such that $c(\varphi_1, \ldots, \varphi_n)$ is a sentence if $\varphi_1, \ldots, \varphi_n$ are. c is *compositional* if the truth or falsity of $c(\varphi_1, \ldots, \varphi_n)$ relative to v only depends on the truth or falsity of $\varphi_1, \ldots, \varphi_n$ relative to v. Assuming coherence of all sentences, including those of the form $c(\varphi_1, \ldots, \varphi_n)$, this means that c is compositional iff there exists a function f_c from $\{1, 0, \wr\}^n$ into $\{1, 0, \wr\}$ such that

$$[\![c(\varphi_1, \ldots, \varphi_n)]\!]_v = f_c([\![\varphi_1]\!]_v, \ldots, [\![\varphi_n]\!]_v)$$

for all v and $\varphi_1, \ldots, \varphi_n$. In other words, a connective of a coherent language is compositional iff its semantical behavior can be specified by a truth table in $1, 0, \wr$. In the following we look at alternative semantical clauses for $\neg$ and $\vee$ that are compositional and yield a coherent truth definition. Such alternative clauses for $\neg$ and $\vee$ correspond to functions from $\{1, 0, \wr\}$ into itself, and from $\{1, 0, \wr\}^2$ into $\{1, 0, \wr\}$ respectively. There are 3^3 and 3^9 such functions, but most of these are without interest if we apply the two criteria of reliability and determinability to the resulting truth definitions. (We recall that a truth definition $\models_x$ is reliable if $u \models_x \varphi^+$ yields $v \models_2 \varphi^+$ and $u \models_x \varphi^-$ yields $v \models_2 \varphi^-$ for all u, φ and completions v of u, and determinable if either $v \models_x \varphi^+$ or $v \models_x \varphi^-$ holds when v is complete.) First observe that under these conditions there is *no* alternative compositional semantical clause for $\neg$. To see this, first suppose v is complete, $v \models \varphi^+$ and $v \models \neg\varphi^-$. Trivially, such v and φ exist. By reliability and determinability of $\models_x$ we must also have $v \models_x \varphi^+$ and $v \models_x \neg\varphi^-$. Assuming that the semantical clauses for $\neg$, relative to $\models_x$, are compositional, we see that the corresponding truth table would have to be consistent with

φ	$\neg\varphi$
1	0
0	
$\wr$	

By an analogous argument, it *also* has to be consistent with

φ	$\neg\varphi$
1	
0	1
$\wr$	

Finally suppose $S \notin v^+$ and $S \notin v^-$. Clearly, such S and v exist. If $\models_x$ is reliable, then

$$
\begin{aligned}
&v \not\models_x S^+\\
&v \not\models_x S^-\\
&v \not\models_x \neg S^+\\
&v \not\models_x \neg S^-
\end{aligned}
$$

Hence the function corresponding to $\neg$ maps $\wr$ to $\wr$, so the truth table would also have to be consistent with

φ	$\neg\varphi$
1	
0	
$\wr$	$\wr$

Putting the three results together, we see that only one truth table is possible

φ	$\neg\varphi$
1	0
0	1
$\wr$	$\wr$

Applying the same type of reasoning to $\vee$, we see that any compositional, determinable and reliable truth definition will have a truth table for $\vee$ consistent with the following:

φ	ψ	$(\varphi \vee \psi)$
1	1	1
1	0	1
1	$\wr$	$\{1, \wr\}$
0	1	1
0	0	0
0	$\wr$	$\wr$
$\wr$	1	$\{1, \wr\}$
$\wr$	0	$\wr$
$\wr$	$\wr$	$\wr$

The set $\{1, \wr\}$ indicates that both 1 and $\wr$ are possible in those lines, but not 0. If we add commutativity, then we see that there is just one alternative to the truth table of the strong Kleene schema. This is known as the *Bochvar* or *weak Kleene* schema:

φ	ψ	$(\varphi \vee \psi)$
1	1	1
1	0	1
1	$\wr$	$\wr$
0	1	1
0	0	0
0	$\wr$	$\wr$
$\wr$	1	$\wr$
$\wr$	0	$\wr$
$\wr$	$\wr$	$\wr$

The strong and weak Kleene truth schemas give rise to alternative truth definitions $\models_3$ and $\models_w$ for $\mathbf{l}$. For many purposes it is more practical to carry out a comparison between the two in the context of a single truth definition for an extended language $\mathbf{l}_{\check{\vee}}$ with two symbols '$\vee$' and '$\check{\vee}$' for strong and weak disjunction. And in fact it is not even necessary to extend $\mathbf{l}$ with a new primitive symbol to this end; relative to the truth definition $\models_3$ it is possible to *define* $\check{\vee}$ from $\vee$ and $\neg$:

$$\models_3 (\varphi \check{\vee} \psi) \rightleftharpoons ((\varphi \vee \psi) \wedge (\varphi \vee \neg\varphi) \wedge (\psi \vee \neg\psi)).$$

From what we have already established about the behavior of $\mathbf{l}$ relative to $\models_3$, it now immediately follows that the weak Kleene truth definition $\models_w$ for $\mathbf{l}$ is coherent, determinable and persistent. We also observe that $\check{\vee}$ is logically weak in the same way as $\vee$. For instance it is *not* the case that for all φ, $\models_3 (\varphi \check{\vee} \neg\varphi)$, or that for all φ and ψ, $\models_3 (\varphi \wedge \psi) \check{\vee} (\varphi \wedge \neg\psi)$ follows from $\models_3 \varphi$. Later on we consider strengthenings of $\models_3$ that are coherent, determinable, reliable (and persistent) and closed under classical consequence. From the above discussion it follows that we cannot expect these truth definitions to be compositional in $\neg$ and $\vee$.

1.3 Encoding in Classical Logic

We have described a 1-1 correspondence between classical models and complete models of the partial format. Gilmore (1974) and Feferman (1984) describe another, more general correspondence between classical models and the full class of partial models. Coupled with an analogous translation procedure for sentences, this provides a very powerful tool which enables us to reduce certain questions about the partial logic to questions about classical logic.

In order to spell out the details, we need a new notion. A *generalized* model is any triple $\langle \rho_v, v^+, v^- \rangle$, where v^+ and v^- are subsets of ρ. Hence all models are generalized models, but to this latter class are admitted also those triples $\langle \rho_v, v^+, v^- \rangle$ for which $v^+ \cap v^- \neq \emptyset$. '$v$' will be used to range over generalized as well as proper models, but only when explicitly indicated. Hence 'for all v' will ordinarily mean 'for all proper models v'. The motivation for studying generalized models is purely technical; no particular interpretation is intended.

We keep the definition of $\models$ unchanged, so now there will be cases where a sentence symbol is both true and false. And given the inductive clauses of $\models$, this predicament will spread upwards to other sentences as well. Hence the resulting logic has *four* "truth values," where the fourth corresponds to *both* true and false. This motivates the subscript '4' below.

In cases where we allow for the possibility of v being a generalized model, we write $v \models_4 \phi$ to indicate this. $\Gamma \models_4 \phi$ holds iff $v \models_4 \phi$ for every generalized model v such that $v \models_4 \phi_0$ for every $\phi_0 \in \Gamma$.

Generalized models are interesting mainly because of a very direct correspondence with classical models. Given a similarity type ρ, let the similarity type ρ^* be the Cartesian product $\rho \times \{+, -\}$. We write 'S^+' and 'S^-' for $\langle S, + \rangle$ and $\langle S, - \rangle$. Here $+$ and $-$ are two distinct objects. A classical model for ρ^* corresponds to a subset of ρ^* itself. Now there is a 1-1 correspondence between generalized models v for ρ and classical models v^* for ρ^*, given by

$$S^+ \in v^* \quad \text{iff} \quad S \in v^+$$

$$S^- \in v^* \quad \text{iff} \quad S \in v^-.$$

For an arbitrary classical model v^* for ρ^*, $\{S \mid S^+ \in v^*\}$ and $\{S \mid S^- \in v^*\}$ may very well overlap. Hence so may the corresponding v^+ and v^-. This is why we need to consider the generalized models.

We recall, from the corresponding general definition, that $\mathbf{l}_{\perp,\wedge}[\rho^*]$ is the set of sentences built up from ρ^* using $\top, \neg, \vee$ together with the added primitive connectives $\bot$ and $\wedge$. For any sentence φ of $\mathbf{l}[\rho]$ we define the positive and negative translation φ^* and φ_*, both of which are in $\mathbf{l}_{\bot,\wedge}[\rho^*]$.

- $\top^* = \top$
- $\top_* = \bot$
- $S^* = S^+$
- $S_* = S^-$
- $(\neg\varphi)^* = \varphi_*$
- $(\neg\varphi)_* = \varphi^*$
- $(\varphi \vee \psi)^* = (\varphi^* \vee \psi^*)$
- $(\varphi \vee \psi)_* = (\varphi_* \wedge \psi_*)$

A *positive* sentence is a sentence without occurrences of the negation symbol $\neg$:

Lemma. * *maps* $\mathbf{l}[\rho]$ *onto the set of positive* $\mathbf{l}_{\bot,\wedge}[\rho^*]$ *sentences.*

Proof: It is immediately seen that φ^* and φ_* are positive for all $\varphi \in \mathbf{l}[\rho]$. On the other hand, for positive $\mathbf{l}_{\bot,\wedge}[\rho^*]$ sentences we can define an *inverse* to $*$.

- $\top^{-*} = \top$
- $\bot^{-*} = \neg\top$
- $(S^+)^{-*} = S$
- $(S^-)^{-*} = \neg S$
- $(\varphi \vee \psi)^{-*} = (\varphi^{-*} \vee \psi^{-*})$
- $(\varphi \wedge \psi)^{-*} = \neg(\neg(\varphi^{-*}) \vee \neg(\psi^{-*}))$

A straightforward induction shows that $(\varphi^{-*})^* = \varphi$. □

Note, however, that $(\varphi^*)^{-*}$ could be distinct from φ. For example, $((\neg\neg S)^*)^{-*} = S$.

Let $(\varphi^+)^*$ and $(\varphi^-)^*$ be φ^* and φ_*, and let Γ^* be the set $\{\phi^* \mid \phi \in \Gamma\}$. We observe the following:

Lemma. *For any* φ, ϕ, Γ *and generalized model* v *we have*

(i) $v \models \varphi^+$ *iff* $v^* \models \varphi^*$

(ii) $v \models \varphi^-$ *iff* $v^* \models \varphi_*$

(iii) $\Gamma \models_4 \phi$ *iff* $\Gamma^* \models_2 \phi^*$

Proof: (iii) follows from (i) and the 1–1 correspondence between classical and generalized models. (i) and (ii) follow by a straightforward, simultaneous induction on φ. $\square$

As a first application of these results we derive some substitution theorems from their classical counterparts. The partial structures, together with the strong Kleene truth definition, provide what has been called a "finer grain," which will sometimes distinguish semantically between sentences that are classically equivalent. As a consequence, truth or falsity in the sense of $\models$ is not always preserved under substitution of classically equivalent sentences. For this reason we shall have to study also the positive and strong equivalence relations defined above. We shall prove that positive equivalence is preserved under substitution of positively equivalent subformulas *that occur positively*, and that strong equivalence is preserved under substitution of *any* strongly equivalent subformulas. To express these results we need to define the appropriate substitution functions. Since we want to distinguish between substitution of positive and negative occurrences, we need *two* functions $\varphi(\psi/^+S)$ and $\varphi(\psi/^-S)$ corresponding to substitution for all positive and all negative occurrences, respectively, of the sentence symbol S. However, we *also* want to define a full substitution function. Since there seems to be no elegant way to define a full substitution function from its positive and negative halves, we opt instead for a more powerful tool of simultaneous substitution. Let 'φ', 'ψ', 'χ', range over arbitrary sentences as before; and let 'S' range over sentence symbols. Intuitively, $\varphi(\psi, \chi/S)$ is the result of substituting ψ for all positive occurrences of S in φ; and χ for all negative occurrences:

Definition. *We define $\varphi(\psi, \chi/S)$ by induction on φ as follows:*

- $\top(\psi, \chi/S) = \top$
- $S(\psi, \chi/S) = \psi$
- $T(\psi, \chi/S) = T$ *for* $T \neq S$
- $(\neg\varphi)(\psi, \chi/S) = \neg(\varphi(\chi, \psi/S))$
- $(\varphi_0 \vee \varphi_1)(\psi, \chi/S) = (\varphi_0(\psi, \chi/S) \vee \varphi_1(\psi, \chi/S))$

The next lemma describes the interaction between this substitution function and the translation functions. We let $\varphi(\begin{smallmatrix}\psi & \chi \\ S & T\end{smallmatrix})$ be the result of a uniform, simultaneous substitution of ψ for S and χ for T in φ.

Lemma. $(\varphi(\psi, \chi/S))^* = \varphi^*(\begin{smallmatrix}\psi^* & \chi_* \\ S^+ & S^-\end{smallmatrix})$ *and* $(\varphi(\psi, \chi/S))_* = \varphi_*(\begin{smallmatrix}\chi^* & \psi_* \\ S^+ & S^-\end{smallmatrix})$.

The proof follows a straightforward induction, and is left to the reader. We shall use the following well known theorem from classical propositional calculus:

Theorem. *Let φ be a positive sentence. For any complete model v, if*

$$v \models \psi_0 \supset \psi_1 \quad \textit{and} \quad v \models \chi_0 \supset \chi_1,$$

then

$$v \models \varphi(\begin{smallmatrix}\psi_0 & \chi_0 \\ S & T\end{smallmatrix}) \supset \varphi(\begin{smallmatrix}\psi_1 & \chi_1 \\ S & T\end{smallmatrix}).$$

As a counterpart, we now get the following:

Theorem. *For any model v and* **l** *sentence φ, if*

$$v \models \psi_0 \supset \psi_1 \quad \textit{and} \quad v \models \neg\chi_0 \supset \neg\chi_1,$$

then

$$v \models \varphi(\psi_0, \chi_0/S) \supset \varphi(\psi_1, \chi_1/S).$$

Proof: This follows from the previous theorem, and the equivalences

$$\begin{array}{lll}
v \models \psi_0 \supset \psi_1 & \text{iff} & v^* \models \psi_0^* \supset \psi_1^* \\
v \models \neg\chi_0 \supset \neg\chi_1 & \text{iff} & v^* \models \chi_{0*} \supset \chi_{1*} \\
\multicolumn{3}{l}{v \models \varphi(\psi_0, \chi_0/S) \supset \varphi(\psi_1, \chi_1/S)} \\
 & \text{iff} & v^* \models (\varphi(\psi_0, \chi_0/S))^* \supset (\varphi(\psi_1, \chi_1/S))^* \\
 & \text{iff} & v^* \models \varphi^*(\begin{smallmatrix}\psi_0^* & \chi_{0*} \\ S^+ & S^-\end{smallmatrix}) \supset \varphi^*(\begin{smallmatrix}\psi_1^* & \chi_{1*} \\ S^+ & S^-\end{smallmatrix}).
\end{array}$$ □

Universally quantifying over v, we also get:

Corollary. *If* $\models_3 \psi_0 \supset \psi_1$ *and* $\models_3 \neg\chi_0 \supset \neg\chi_1$*, then*

$$\models_3 \varphi(\psi_0, \chi_0/S) \supset \varphi(\psi_1, \chi_1/S).$$

By symmetry, this also holds with '$\equiv$' substituted for '$\supset$'. If we now set $S = \chi_0 = \chi_1$ and define $\varphi(\psi/^+S)$ as $\varphi(\psi, S/S)$, we immediately get:

Corollary. *If* $\models_3 \psi \equiv \chi$*, then* $\models_3 \varphi(\psi/^+S) \equiv \varphi(\chi/^+S)$.

Defining $\varphi(\psi/S)$ as $\varphi(\psi, \psi/S)$, we also get:

Corollary. *If* $\models_3 \psi \rightleftharpoons \chi$*, then* $\models_3 \varphi(\psi/S) \rightleftharpoons \varphi(\chi/S)$.

Proof: Suppose $\models_3 \psi \rightleftharpoons \chi$. Then

$$\models_3 \psi \equiv \chi \quad \text{and} \quad \models_3 \neg\psi \equiv \neg\chi;$$

hence

$$\models_3 \varphi(\psi, \psi/S) \equiv \varphi(\chi, \chi/S)$$

and

$$\models_3 (\neg\varphi)(\psi, \psi/S)) \equiv (\neg\varphi)(\chi, \chi/S)),$$

and thus

$$\models_3 \varphi(\psi, \psi/S) \rightleftharpoons \varphi(\chi, \chi/S). \qquad \square$$

Despite their status as corollaries we shall call these the *positive* and the *strong* equivalence theorems.

1.4 Extending the Language

In classical logic, the connectives $\top$, $\vee$, $\neg$ of $\mathbf{l}$ constitute a truth-functionally complete set of connectives; every function from $\{1,0\}^n$ into $\{1,0\}$ can be expressed by a sentence in these connectives. This is why in classical logic we often restrict attention to these connectives.

Our partial models, on the other hand, have an associated logic that could be construed as three valued, with *true, false* and *neither* as the three possible "truth values." With this generalization introduced, the set $\{\top, \vee, \neg\}$ is no longer complete; not all functions from $\{1, 0, \wr\}^n$ into $\{1, 0, \wr\}$ can be expressed by a sentence in these connectives.

For any set $\{c_1, \ldots, c_n\}$ of compositional connectives, we may form the language $\mathbf{l}_{c_1,\ldots,c_n}$ where $c_1, \ldots, c_n$ are added to the connectives of $\mathbf{l}$. With such additions we will be able to express more truth functions. We shall see, however, that any such increase in expressibility obtained by adding new compositional propositional connectives will be won at the cost of violating one or more of our three conditions of coherence, determinability and persistence. No addition of compositional propositional connectives can bring to the language any new coherent, determinable and persistent sentence which is not strongly equivalent to a sentence of $\mathbf{l}$.

We have seen that the partial models have an associated logic that can be construed as three valued. Theoretically there is a fourth possibility of simultaneous truth and falsity, and we notice that the logic is three valued rather than four valued only because the connectives $\top$, $\neg$, $\vee$ are "well behaved." And even with these connectives, the possibility of a fourth truth value is realized when the coherence condition on models is not assumed, i.e., when we turn to generalized models. In the resulting more general setting, a compositional connective will not necessarily correspond to a function into $\{1,0,\wr\}$. We have to consider instead the functions from $\{1,0,\wr,\times\}^n$ into $\{1,0,\wr,\times\}$, where $\times$ corresponds to "both true and false,"

$$[\![\varphi]\!]_v = \times \quad \text{iff} \quad v \models \varphi^+ \quad \text{and} \quad v \models \varphi^- .$$

In the general setting of not necessarily coherent languages, a compositional n-ary connective c has an associated function f_c from $\{1,0,\wr,\times\}^n$ into $\{1,0,\wr,\times\}$ such that for all sentences $\varphi_1,\ldots,\varphi_n$ and partial models v

$$[\![c(\varphi_1,\ldots,\varphi_n)]\!]_v = f_c([\![\varphi_1]\!]_v,\ldots,[\![\varphi_n]\!]_v).$$

When such an f_c exists, the truth definition for sentences containing c can be extended to *generalized* partial models, using the same general equation.

In this section we identify a set of compositional connectives that is truth functionally complete with respect to four valued logic in the same way that $\top$, $\vee$ and $\neg$ are complete with respect to two valued logic. Let $\star$ and $\diamond$ be two 0-ary connectives, with the following semantical clauses:

$$v \not\models \star^+ \quad \text{and} \quad v \not\models \star^-$$

$$v \models \diamond^+ \quad \text{and} \quad v \models \diamond^-$$

for all v. Hence $[\![\star]\!]_v = \wr$ and $[\![\diamond]\!]_v = \times$ for all v. We first show that the connectives $\top$, $\neg$, $\vee$, $\sim$, $\star$, $\diamond$ are complete relative to the set of functions from $\{1,0,\wr,\times\}^n$ into $\{1,0,\wr,\times\}$. Thus any sentence of an extension of $\mathbf{l}$ obtained by adding compositional connectives will be strongly equivalent to a sentence of $\mathbf{l}_{\sim,\star,\diamond}$. Following this result, we shall compare the four languages $\mathbf{l}$, $\mathbf{l}_{\sim}$, $\mathbf{l}_{\sim,\star}$ and $\mathbf{l}_{\sim,\star,\diamond}$: Trivial proofs by induction show that $\mathbf{l}_{\sim,\star}$ is coherent, $\mathbf{l}_{\sim}$ is coherent and determinable, and we already know that $\mathbf{l}$ has all three properties. The connective $\sim$ gives rise to non-persistent sentences, $\star$ to non-determinable, and $\diamond$ to non-coherent ones. Indeed, the sentence $\diamond$ is itself non-coherent, $\star$ is itself non-determinable and $\sim S$ is non-persistent for the sentence symbol S. These observations are immediate. We shall prove the much deeper results, however, that allowing $\sim$ adds *nothing but* non-persistent sentences to $\mathbf{l}$, allowing $\star$ adds *nothing but* non-determinable sentences to $\mathbf{l}_{\sim}$, and allowing $\diamond$ adds *nothing but* non-coherent sentences to $\mathbf{l}_{\sim,\star}$, in the sense that every persistent sentence of $\mathbf{l}_{\sim}$ is strongly equivalent to a sentence of $\mathbf{l}$, etc. Hence $\mathbf{l}$ can be viewed as the restriction of the truth-functionally complete language $\mathbf{l}_{\sim,\star,\diamond}$ to its set of "legitimate" sentences, those that satisfy the three conditions of coherence, determinability and persistence.

We shall again make use of the correspondence between generalized and classical models, and we extend the translations φ^* and φ_* to $\mathbf{l}_{\sim,\star,\diamond}$ sentences φ in general. The additional clauses are the following:

$$(\sim\varphi)^* = \neg(\varphi^*) \qquad (\sim\varphi)_* = \varphi^*$$

$$\star^* = \star_* = \bot \qquad \diamond^* = \diamond_* = \top$$

The proof of the next lemma is left to the reader:

Lemma. *For any $\mathbf{l}_{\sim,\star,\diamond}$ sentence φ, $v \models_4 \varphi^+$ iff $v^* \models_2 \varphi^*$, and $v \models_4 \varphi^-$ iff $v^* \models_2 \varphi_*$.*

Since the mapping $v \to v^*$ is onto, we also obtain

Lemma. *For any $\mathbf{l}_{\sim,\star,\diamond}$ sentences φ and ψ, $\models_4 \varphi \equiv \psi$ iff $\models_2 \varphi^* \equiv \psi^*$.*

Before we state the main lemma of this section, we notice that the inverse $^{-*}$ to * can be extended so as to apply to the full set of $\mathbf{l}_{\wedge,\bot}[\rho^*]$ sentences, rather than just the positive ones, by adding the clause

$$(\neg\varphi)^{-*} = \sim(\varphi^{-*}).$$

With this definition, the equality $(\varphi^{-*})^* = \varphi$ still holds. (But note that *, as defined on $\mathbf{l}_{\sim,\star,\diamond}$, is even less injective than on $\mathbf{l}$, and the general equality $(\varphi^*)^{-*} = \varphi$ still has counterexamples.)

It is seen that if φ is an $\mathbf{l}_{\wedge,\bot}[\rho^*]$ sentence, then φ^{-*} is an $\mathbf{l}_{\sim}[\rho]$ sentence, and hence that * maps $\mathbf{l}_{\sim}[\rho]$ onto $\mathbf{l}_{\wedge,\bot}[\rho^*]$.

We use the following functional completeness result for $\mathbf{l}$ relative to classical propositional logic. The result holds for finite similarity types ρ.

Theorem. *Let K_ρ be the set of classical ρ models. For any function f from K_ρ into $\{0,1\}$ there is an $\mathbf{l}[\rho]$ sentence φ such that $[\![\varphi]\!]_v = f(v)$ for all $v \in K_\rho$.*

From this, we obtain a similar result about $\mathbf{l}_{\sim,\star,\diamond}$ relative to generalized partial models.

Theorem. *Let G_ρ be the set of generalized ρ models. For any function f from G_ρ into $\{0,1,\wr,\times\}$ there is an $\mathbf{l}_{\sim,\star,\diamond}[\rho]$ sentence φ such that $[\![\varphi]\!]_v = f(v)$ for all $v \in G_\rho$.*

Proof: For any such function f, let f^* and f_* be the functions from K_{ρ^*} into $\{0,1\}$, such that

$$f^*(v^*) = 1 \quad \text{iff} \quad f(v) \in \{1,\times\}$$

$$f_*(v^*) = 1 \quad \text{iff} \quad f(v) \in \{0,\times\}.$$

By the above theorem there are $\mathbf{l}[\rho^*]$ sentences χ^p and χ^n such that for any $v \in G_\rho$,

$$[\![\chi^p]\!]_{v^*} = f^*(v^*)$$

$$[\![\chi^n]\!]_{v^*} = f_*(v^*).$$

Hence

$$[\![\chi^{p-*}]\!]_v \in \{1, \times\} \quad \text{iff} \quad f(v) \in \{1, \times\}$$

$$[\![\chi^{n-*}]\!]_v \in \{1, \times\} \quad \text{iff} \quad f(v) \in \{0, \times\}$$

Now observe that

$$[\![((\chi^{p-*} \wedge \diamond) \vee (\neg(\chi^{n-*}) \wedge \star))]\!]_v \in \{1, \times\} \quad \text{iff} \quad [\![\chi^{p-*}]\!]_v \in \{1, \times\}$$

$$[\![((\chi^{p-*} \wedge \diamond) \vee (\neg(\chi^{n-*}) \wedge \star))]\!]_v \in \{0, \times\} \quad \text{iff} \quad [\![\chi^{n-*}]\!]_v \in \{1, \times\}$$

Consequently

$$[\![((\chi^{p-*} \wedge \diamond) \vee (\neg(\chi^{n-*}) \wedge \star))]\!]_v = f(v). \qquad \square$$

As a trivial consequence, we also get the following.

Theorem. *Let V_ρ be the set of partial ρ models. For any function f from V_ρ into $\{0, 1, \wr, \times\}$ there is an $\mathbf{l}_{\sim,\star,\diamond}[\rho]$ sentence φ such that $[\![\varphi]\!]_v = f(v)$ for all $v \in V_\rho$.*

This result tells us that $\mathbf{l}_{\sim,\star,\diamond}$ has maximal expressive power relative to the partial models. To make this precise, we introduce the notion of *languages* in general. Rather than to define this notion, we stipulate some necessary conditions. If $\mathbf{l}^x$ is a language for propositional logic, then $\mathbf{l}^x$ identifies for every ρ a set $\mathbf{l}^x[\rho]$ of *sentences*. Moreover, $\mathbf{l}^x$ identifies for every ρ a semantical relation $\models_{\mathbf{l}^x}$ between ρ models and *signed* $\mathbf{l}^x[\rho]$ sentences ϕ. The following conditions will not play a crucial part in any proof of this chapter, but they seem natural. More could be added.

(i) If $\rho \subseteq \rho'$, then $\mathbf{l}^x[\rho] \subseteq \mathbf{l}^x[\rho']$.

(ii) If v is a ρ' model, $\rho \subseteq \rho'$ and $\phi \in \mathbf{l}^x[\rho]$, then $v \models_{\mathbf{l}^x} \phi$ iff $v \uparrow \rho \models_{\mathbf{l}^x} \phi$.

Theorem. *Let $\mathbf{l}^x$ be a language and ρ a similarity type. Then for every $\mathbf{l}^x[\rho]$ sentence ψ there is an $\mathbf{l}_{\sim,\star,\diamond}[\rho]$ sentence φ such that $v \models \varphi^\odot$ iff $v \models_{\mathbf{l}^x} \psi^\odot$.*

Proof: This is obvious, since every $\psi \in \mathbf{l}^x[\rho]$ corresponds to a function from V_ρ into $\{0, 1, \wr, \times\}$. $\square$

As a special case, we also get:

Theorem. *Let C be a set of compositional connectives. Then for any $\mathbf{l}_C[\rho]$ sentence ψ there is an $\mathbf{l}_{\sim,\star,\diamond}[\rho]$ sentence φ such that $\models_3 \varphi \rightleftharpoons \psi$.*

We next show that every coherent sentence of $\mathbf{l}_{\sim,\star,\diamond}$ is strongly equivalent to an $\mathbf{l}_{\sim,\star}$ sentence, and that every determinable sentence of $\mathbf{l}_{\sim,\star}$ is strongly equivalent to a sentence of $\mathbf{l}_{\sim}$. The following lemma takes us part of the way. Note that this result holds for *all* $\mathbf{l}_{\sim,\star,\diamond}$ sentences, regardless of whether they are coherent, or determinable.

Lemma. *For every $\mathbf{l}_{\sim,\star,\diamond}[\rho]$ sentence φ there is an $\mathbf{l}_{\sim}[\rho]$ sentence ψ such that $\models_4 \varphi \equiv \psi$.*

Proof: $v \models_4 \varphi$ iff $v^* \models_2 \varphi^*$ iff $v \models_4 (\varphi^*)^{-*}$. Hence $\models_4 \varphi \equiv (\varphi^*)^{-*}$. By definition $(\varphi^*)^{-*}$ has no occurrences of $\diamond$ or $\star$. □

From this we obtain a useful normal form result:

Lemma. *Every $\mathbf{l}_{\sim,\star,\diamond}[\rho]$ sentence is strongly equivalent to an $\mathbf{l}_{\sim,\star,\diamond}[\rho]$ sentence of the form $((\varphi^p \wedge \diamond) \vee (\varphi^n \wedge \star))$, where φ^p and φ^n contain no occurrences of $\diamond$ or $\star$.*

Proof: Let φ be any $\mathbf{l}_{\sim,\star,\diamond}[\rho]$ sentence. By the previous lemma there exist $\mathbf{l}_{\sim}[\rho]$ sentences φ^p and φ^n such that

$$\models_4 \varphi \equiv \varphi^p \quad \text{and} \quad \models_4 \neg\varphi \equiv \neg\varphi^n.$$

The following equivalence is now easily checked:

$$\models_4 \varphi \rightleftharpoons ((\varphi^p \wedge \diamond) \vee (\varphi^n \wedge \star)).$$ □

We recall that *coherence* refers to the behavior of sentences in proper partial models. In the proof of the next theorem, note that strong equivalence is shown only in proper partial models:

Theorem. *Every coherent $\mathbf{l}_{\sim,\star,\diamond}[\rho]$ sentence is strongly equivalent to an $\mathbf{l}_{\sim,\star}[\rho]$ sentence.*

Proof: Let φ be a coherent $\mathbf{l}_{\sim,\star,\diamond}[\rho]$ sentence. By the above lemma, we can assume that φ is of the form $(\varphi^p \wedge \diamond) \vee (\varphi^n \wedge \star)$, where φ^p and φ^n are $\mathbf{l}_{\sim}$ sentences. Since φ is coherent, it follows that $\models_3 \neg\varphi^n \supset \neg{\sim}{\sim}\varphi^p$. And this implies the equivalence

$$\models_3 ((\varphi^p \wedge \diamond) \vee (\varphi^n \wedge \star)) \rightleftharpoons ({\sim}{\sim}\varphi^p \vee (\varphi^n \wedge \star)).$$ □

Since every $\mathbf{l}_{\sim,\star}[\rho]$ sentence is coherent, the above proof also gives us the following normal form lemma:

Lemma. *Every $\mathbf{l}_{\sim,\star}[\rho]$ sentence is strongly equivalent to an $\mathbf{l}_{\sim,\star}[\rho]$ sentence of the form $\varphi \vee (\psi \wedge \star)$, where φ and ψ contain no occurrences of $\star$.*

For any given (finite) similarity type ρ, let *compl* be the sentence

$$\bigwedge_{S \in \rho}(S \vee \neg S).$$

Now the following lemma is all we need:

Lemma. *If the* $\mathbf{l}_{\sim,\star}$ *sentence* $(\varphi \vee (\psi \wedge \star))$ *is determinable, then* $\models_3 (\psi \wedge \mathit{compl}) \supset \varphi$.

Proof: Suppose $v \models (\psi \wedge \mathit{compl})^+$ and $v \not\models \varphi^+$. By the first assumption v is complete, and by the second $v \not\models (\varphi \vee (\psi \wedge \star))^+$. Since ψ is an $\mathbf{l}_{\sim,\star}$ sentence and thus coherent, $v \not\models \psi^-$ and hence $v \not\models (\varphi \vee (\psi \wedge \star))^-$. But this is impossible since $\varphi \vee (\psi \wedge \star)$ is determinable. □

Theorem. *Every determinable* $\mathbf{l}_{\sim,\star}[\rho]$ *sentence is strongly equivalent to an* $\mathbf{l}_{\sim}[\rho]$ *sentence.*

Proof: Let χ be a determinable $\mathbf{l}_{\sim,\star}$ sentence. We may assume that χ is of the form $(\varphi \vee (\psi \wedge \star))$, where neither φ nor ψ contains occurrences of $\star$. Since χ is determinable, by the lemma above we obtain:

$$\models_3 (\varphi \vee (\psi \wedge \star)) \rightleftharpoons (\varphi \vee (\psi \wedge \mathit{compl}))$$

Clearly, $(\varphi \vee (\psi \wedge \mathit{compl}))$ is an $\mathbf{l}_{\sim}$ sentence. □

We proceed to show that every persistent $\mathbf{l}_{\sim}$ sentence is strongly equivalent to an $\mathbf{l}$ sentence. We use the following auxiliary notions. A sentence φ is *truth-persistent* if for all u and v, if $u \ll v$ and $u \models \varphi^+$, then $v \models \varphi^+$. φ is *falsity-persistent* if for all u and v if $u \ll v$ and $u \models \varphi^-$, then $v \models \varphi^-$. Truth-persistence and falsity-persistence relative to generalized structures are defined similarly.

Hence a sentence is persistent iff it is both truth-persistent and falsity-persistent. In characterizing the persistent sentences, it will be helpful first to characterize the truth-persistent and falsity-persistent sentences separately. With little effort, these characterizations can be obtained using results from classical logic and the translation procedures previously considered. But this, also, is how far the translation procedures take us. The function * preserves truth and the dual function $_*$ does, in a sense, preserve falsity, but neither preserves both. The translation procedures can help us to analyze the truth behavior or the falsity behavior of a sentence, but not both at the same time. We shall later patch the two separate characterization results together, using the "cut-and-glue" theorem, which we prove towards the end of this chapter. We introduce the following notion from classical logic:

A sentence φ is *increasing* if for all *classical* models u and v, if $u \subseteq v$, then $u \models \varphi$ implies $v \models \varphi$. We have the following:

Lemma. *Let φ be an $\mathbf{l}_{\sim}$ sentence. φ is truth-persistent with respect to generalized models iff φ^* is increasing.*

Proof: This is so because $u \ll v$ iff $u^* \subseteq v^*$, and because the mapping $v \to v^*$ is *onto.* □

The following theorem from classical logic is now relevant.

Theorem. *An $\mathbf{l}_{\wedge,\perp}[\rho]$ sentence is increasing iff it is classically equivalent to a positive $\mathbf{l}_{\wedge,\perp}[\rho]$ sentence.*

For a proof, see Chang & Keisler (1977), pp. 13-14. The theorem proved there is a slightly different result from the above, since it is stated for sentences without occurrences of $\top$ and $\perp$, and therefore takes a slightly different form. The present version is easily deduced, however, the crucial step being that the sentence $(S^\top \wedge (S^\perp \vee \varphi\left(\begin{smallmatrix} S^\top & S^\perp \\ \top & \perp \end{smallmatrix}\right)))$ is increasing if φ is, where $S^\top$ and $S^\perp$ are sentence symbols without occurrences in φ. Hence we now get the following:

Lemma. *An $\mathbf{l}_{\sim}[\rho]$ sentence φ is truth-persistent with respect to generalized models iff there exists an $\mathbf{l}[\rho]$ sentence ψ such that $\models_4 \varphi \equiv \psi$.*

Proof: Equivalence is already established between consecutive items below:

1. φ is truth-persistent with respect to generalized models.
2. φ^* is increasing.
3. There exists a positive $\mathbf{l}_{\wedge,\perp}[\rho^*]$ sentence χ such that $\models_2 \varphi^* \equiv \chi$.
4. There exists an $\mathbf{l}[\rho]$ sentence ψ such that $\models_2 \varphi^* \equiv \psi^*$.
5. There exists an $\mathbf{l}[\rho]$ sentence ψ such that $\models_4 \varphi \equiv \psi$. □

However, we are primarily interested in sentences truth-persistent with respect to proper models. We shall obtain a similar theorem for such sentences, but first we need a lemma. For any given (finite) similarity type ρ, let *contr* be the sentence

$$\bigvee\!\!\!\!\bigvee_{S\in\rho}(S \wedge \neg S).$$

Note that *contr* is strongly equivalent to $\neg$*compl.*

Lemma. *If φ is truth-persistent, then $(\varphi \vee contr)$ is truth-persistent with respect to generalized models.*

Proof: Suppose φ is truth-persistent, u and v are generalized models such that $u \ll v$, and suppose $u \models (\varphi \vee contr)$. We must show that $v \models (\varphi \vee contr)$.

- If $v \models contr$, then $v \models (\varphi \vee contr)$.
- If $v \not\models contr$, then both u and v are proper. Hence $u \models \varphi$. By truth-persistence of φ with respect to proper models, $v \models \varphi$ and $v \models (\varphi \vee contr)$. □

Theorem. *An* $\mathbf{l}_{\sim}[\rho]$ *sentence* φ *is truth-persistent iff there exists an* $\mathbf{l}[\rho]$ *sentence* ψ *such that* $\models_3 \varphi \equiv \psi$.

Proof: If φ is truth-persistent, then $(\varphi \vee contr)$ is truth-persistent with respect to generalized models. Hence there is an $\mathbf{l}[\rho]$ sentence ψ such that $\models_4 (\varphi \vee contr) \equiv \psi$. Hence $\models_3 \varphi \equiv \psi$. □

Corollary. *An* $\mathbf{l}_{\sim}[\rho]$ *sentence* φ *is falsity-persistent iff there exists an* $\mathbf{l}[\rho]$ *sentence* χ *such that* $\models_3 \neg\varphi \equiv \neg\chi$.

Combining the separate truth- and falsity-persistence characterizations, we obtain the following characterization of full persistence:

Proposition. *An* $\mathbf{l}_{\sim}[\rho]$ *sentence* φ *is persistent iff there exist* $\mathbf{l}[\rho]$ *sentences* ψ *and* χ *such that* $\models_3 \varphi \equiv \psi$ *and* $\models_3 \neg\varphi \equiv \neg\chi$.

Later we shall improve on this result, and show that an $\mathbf{l}_{\sim}[\rho]$ sentence is persistent iff it is strongly equivalent to an $\mathbf{l}[\rho]$ sentence. Hence the more general claim will also follow:

Theorem. *Let* $\mathbf{l}^x$ *be a language. An* $\mathbf{l}^x[\rho]$ *sentence is coherent, determinable and persistent iff it is strongly equivalent to an* $\mathbf{l}[\rho]$ *sentence.*

But this result is not within reach for a while yet. What we need is the "cut-and-glue" theorem. The latter is an interesting result in its own right, and it is discussed and proved over the next several pages. It should be mentioned that this chosen route to the above coherence-determinability-persistence characterization result, *via* the "cut-and-glue" theorem, is certainly not the shortest or simplest. In fact a straightforward truth table argument is sufficient, cf. van Benthem (1988). But unlike the present strategy, such an argument does not generalize to predicate logic.

We conclude this section with a few observations about some non-compositional operators:

$$u \models \Box\varphi^{\odot} \quad \text{iff} \quad (v \models \varphi^{\odot} \text{ for all completions } v \text{ of } u).$$

We observe that $\Box\varphi$ is persistent even if φ itself is not. Let $\uparrow(\varphi)$ and $\downarrow(\varphi)$ be

$$\varphi \vee \Box{\sim}\neg\varphi \quad \text{and} \quad \varphi \wedge \neg\Box{\sim}\varphi$$

respectively. For persistent φ, we see that whenever φ is defined, then $\uparrow(\varphi)$ and $\downarrow(\varphi)$ both have the same truth value as φ. Otherwise $\uparrow(\varphi)$ and $\downarrow(\varphi)$ are true and false, respectively, "as often as possible." If φ is persistent, then $\uparrow(\varphi)$ and $\downarrow(\varphi)$ are combinations in $\vee, \wedge, \neg$ of persistent sentences, and hence themselves persistent. Moreover, $\uparrow(\varphi)$ and $\downarrow(\varphi)$ are determinable for any φ, and coherent if φ is. Hence if φ is coherent and persistent, then $\uparrow(\varphi)$ and $\downarrow(\varphi)$ are coherent, determinable and persistent.

The following strong equivalence is easily checked.

$$\models_3 \varphi \rightleftharpoons (\downarrow(\varphi) \vee (\uparrow(\varphi) \wedge \star)).$$

Hence from the above coherence-determinability-persistence characterization result, we obtain the following coherence-persistence characterization result, which is due to S. Blamey.

Theorem. *Let* $\mathbf{l}^x$ *be a language. An* $\mathbf{l}^x[\rho]$ *sentence is coherent and persistent iff it is strongly equivalent to an* $\mathbf{l}_\star[\rho]$ *sentence.*

1.5 The "Cut-and-Glue" Theorem: Statement

One piece is missing in our proof of the full coherence-determinability-persistence characterization theorem. We must show that whenever a pair of $\mathbf{l}$ sentences ψ and χ exists, satisfying

$$\models_3 \varphi \equiv \psi \quad \text{and} \quad \models_3 \neg\varphi \equiv \neg\chi,$$

where φ is coherent, determinable and persistent, then also a single $\mathbf{l}$ sentence exists that fills the rôle of both, i.e., $\models_3 \varphi \rightleftharpoons \mu$ for some $\mathbf{l}$ sentence μ.

When both φ and χ are coherent and determinable, then $\models_3 \neg\varphi \equiv \neg\chi$ implies $\models_2 \varphi \equiv \chi$. Hence with the appropriate assumptions,

$$\models_3 \varphi \equiv \psi \quad \text{and} \quad \models_3 \neg\varphi \equiv \neg\chi$$

implies

$$\models_2 \psi \equiv \chi,$$

and the existence of the μ above follows immediately from the "cut-and-glue" theorem, which states that for every two classically equivalent $\mathbf{l}$ sentences there is a third $\mathbf{l}$ sentence that is positively equivalent to the first, and negatively equivalent to the second.

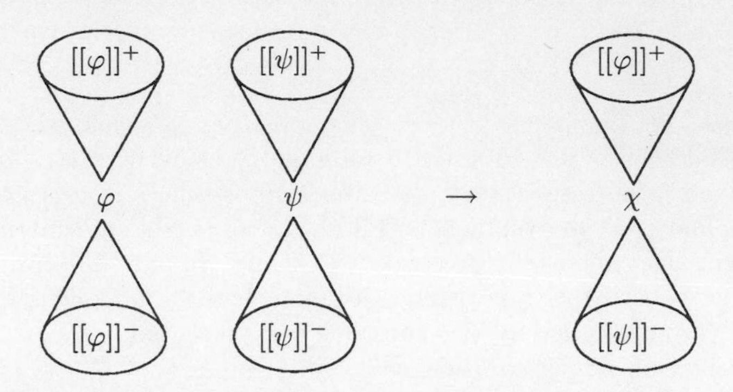

The above figure shows the motivation for the name of this theorem. If $[\![\varphi]\!]^+$ and $[\![\varphi]\!]^-$ are defined as $\{v \mid v \models \varphi^+\}$ and $\{v \mid v \models \varphi^+\}$ then

$$\models_3 \varphi \equiv \psi \quad \text{iff} \quad [\![\varphi]\!]^+ = [\![\psi]\!]^+.$$

The full semantical behavior of φ is determined by the pair

$$[\![\varphi]\!] = \langle [\![\varphi]\!]^+, [\![\varphi]\!]^- \rangle.$$

The "cut-and-glue" theorem says that under certain conditions we can "cut" such pairs in two and "glue" the parts together. We prove this theorem towards the end of the chapter; most of the intermediate material lead up to this result, or represent various natural digressions along the way. In this section we make some remarks on what this theorem really amounts to, and why it is not the rather trivial result that it may seem to be at first glance.

For an extension of **l** containing a connective | such that

$$\models_3 (\varphi \mid \psi) \equiv \varphi \quad \text{and} \quad \models_3 \neg(\varphi \mid \psi) \equiv \neg\psi,$$

the "cut-and-glue" theorem would of course be trivial. Such a connective is not definable in **l** itself, however, since it gives rise to non-coherent as well as non-determinable sentences. But we would still get an easy proof of the "cut-and-glue" theorem if we could find a compositional connective $\circ$ that were both definable in **l**, and at the same time would act as an approximation to | on classically equivalent sentences, i.e., such that for **l** sentences φ and ψ,

$$\models_3 (\varphi \mid \psi) \rightleftharpoons (\varphi \circ \psi) \quad \text{whenever} \quad \models_2 \varphi \equiv \psi.$$

We show that no such $\circ$ exists. Suppose otherwise. Then consider the two pairs of classically equivalent **l** sentences $\top$ and $(S \vee \neg S)$, and $(S \wedge \neg S)$ and

$\bot$. By assumption $\top \circ (S \vee \neg S)$ is true iff $\top$ is, and $(S \wedge \neg S) \circ \bot$ is false iff $\bot$ is. The truth function $f_\circ$ associated with the connective $\circ$ then satisfies

$$f_\circ(1, \imath) = 1 \quad \text{and} \quad f_\circ(\imath, 0) = 0.$$

But now we get a contradiction. For although we have only assumed $\circ$ to behave like $|$ on pairs of classically equivalent sentences, $\circ$ is still assumed to be a compositional connective, and $(\varphi \circ \psi)$ is a sentence for any **l** sentences φ and ψ. Hence $f_\circ(1, 0)$ is equal to 1, 0 or $\imath$. But any choice here gives us a non-monotonic function $f_\circ$, and a non-persistent sentence $\varphi \circ \psi$ for some φ and ψ.

We conclude that this simplest conceivable strategy to prove the "cut-and-glue" theorem cannot work; some more elaborate transformation is required. In the case of predicate logic we shall prove that no recursive function exists that maps all pairs of classically equivalent sentences to witness sentences of the form needed. This proof uses the closely related notion of *relative saturants.*

The relative saturation theorem for propositional logic states that for every two **l** sentences φ and ψ such that $\models_2 \varphi \supset \psi$ there exists an **l** sentence χ such that $\models_3 \varphi \supset \chi$ and $\models_3 \neg\chi \equiv \neg\psi$. Hence χ is (deductively) *saturated* relative to φ, in that $\models_2 \varphi \supset \chi$ implies $\models_3 \varphi \supset \chi$. Alternatively stated, the relative saturation theorem says that for any two sentences φ and ψ there exists a sentence χ, which is saturated relative to φ, and negatively equivalent to ψ. Before we proceed, we indicate the connection between the relative saturation theorem and the "cut-and-glue" theorem. Given that both of these theorems hold, the significance of the results below lies in the simplicity of the proofs, and the fact that so few assumptions are needed.

Proposition. *If the "cut-and-glue" theorem holds, then the relative saturation theorem holds.*

Proof: We assume the "cut-and-glue" theorem. Suppose $\models_2 \varphi \supset \psi$. Then $\models_2 \psi \equiv (\varphi \vee \psi)$, and a χ exists such that $\models_3 \neg\psi \equiv \neg\chi$ and $\models_3 (\varphi \vee \psi) \equiv \chi$. Hence also $\models_3 \varphi \supset \chi$. □

Proposition. *If the the relative saturation theorem holds, then the "cut-and-glue" theorem holds.*

Proof: We assume the relative saturation theorem. Suppose $\models_2 \varphi \equiv \psi$. Then also $\models_2 \varphi \supset \psi$, and hence a χ exists such that $\models_3 \varphi \supset \chi$ and $\models_3 \neg\psi \equiv \neg\chi$. Simple calculations now show that

$$\models_3 \varphi \equiv (\chi \wedge (\varphi \vee \neg\varphi))$$

$$\models_3 \neg\psi \equiv \neg(\chi \wedge (\varphi \vee \neg\varphi))$$

To see that the first of these holds, observe that $\models_2 \varphi \equiv \psi$ and $\models_3 \neg\psi \equiv \neg\chi$ together imply $\models_2 \varphi \equiv \chi$. Hence by persistence of φ and χ, also $\models_3 \chi \supset \sim\neg\varphi$. □

The relative saturation theorem for predicate logic is in a certain sense a maximal result; no obvious strengthening seems to hold. But in the case of propositional logic we shall prove a uniform version where the saturant χ is independent of φ. In other words, we show that every **1** sentence is negatively equivalent to some *saturated sentence*, a sentence χ that satisfies the implication

$$\models_2 \varphi \supset \chi \quad \text{implies} \quad \models_3 \varphi \supset \chi$$

for all **1** sentences φ.

This notion of saturated sentences leads us to a study of alternative truth definitions, in particular the *supervaluation truth definition* $\models_\Box$, introduced by Bas van Fraassen. $\models_\Box$ is defined by the equation

$$v \models_\Box \phi \quad \text{iff} \quad (v' \models \phi \text{ for all completions } v' \text{ of } v),$$

where 'ϕ' ranges over signed sentences. As we shall see, an **1** sentence

$$v \models_\Box \varphi \quad \text{iff} \quad v \models \varphi.$$

Using this characterization, we shall deduce the saturation theorem for **1** from results about the relation between $\models_\Box$ and $\models$. In predicate logic the above characterization of saturatedness breaks down, and we shall have to turn instead to yet other truth definitions.

1.6 Alternative Truth Definitions

We have defined a truth definition $\models_x$ to be *reliable* if $v \models_x \phi$ implies $v' \models_2 \phi$ for any completion v' of v. The supervaluation truth definition was defined above. Defining a truth definition $\models_x$ to be *at least as strong as* another truth definition $\models_y$, written '$\models_y \leq \models_x$', iff $v \models_y \phi$ implies $v \models_x \phi$ for all v and ϕ, we see that a truth definition is reliable iff $\models_\Box$ is at least as strong as this truth definition.

We shall prove two characterizations of $\models_\Box$ that hold for propositional calculus but not for predicate logic. When these two characterizations are established, we will also have gained enough insight to prove the saturation theorem and the "cut-and-glue" theorem.

Closure under tautological inference is a chief feature of the supervaluation truth definition; it follows immediately from the definition of $\models_\Box$ that if $\Gamma \models_2 \phi$ and $v \models_\Box \phi_0$ for all $\phi_0 \in \Gamma$, then $v \models_\Box \phi$. It turns out that in propositional calculus this property characterizes $\models_\Box$ among reliable truth definitions at least as strong as $\models$. To make this precise, we define $\models^{scl}$, the *syntactic closure* of $\models$, to be the smallest relation satisfying the following:

1. If $v \models \phi$, then $v \models^{scl} \phi$.

2. If $\Gamma \models_2 \phi$, $\phi \in \mathbf{l}[\rho_v]$ and $v \models^{scl} \phi_0$ for all $\phi_0 \in \Gamma$, then $v \models^{scl} \phi$.

Since $v \models \phi$ only if $\phi \in \mathbf{l}[\rho_v]$, it follows that similarly $v \models^{scl} \phi$ only if $\phi \in \mathbf{l}[\rho_v]$. Clearly $\models^{scl}$ is a well defined relation, but note that $v \models^{scl} \phi$ is potentially dependent on the similarity type ρ_v. Let ρ_φ be the set of sentence symbols occurring in φ. For all that is established so far, we could have $v \models \psi$ and $\psi \models_2 \varphi$ for some ψ of a larger similarity type, but not $v \models \psi'$ and $\psi' \models_2 \varphi$ for any ρ_φ sentence ψ'. Hence for a ρ sentence φ and ρ' model v, where $\rho \subseteq \rho'$, we could have $v \models^{scl} \varphi$ but not $v \uparrow \rho \models^{scl} \varphi$. As it turns out, however, there is no such dependence. This will follow from the result that $\models^{scl} = \models_\Box$.

We shall also turn to the question of *why* the strong Kleene schema gives us a truth definition that is weaker than $\models_\Box$. In a previous section we saw that this relative weakness was a fate shared with all truth definitions that are compositional, reliable and determinable. Following up this insight, we shall experiment with a variant of $\models$ with defining clauses that represent a minimal deviation from compositionality. Rather than evaluate subformulas in isolation, we shall use a more general format which allows for the simultaneous evaluation of finitely many signed sentences, understood disjunctively. The conditions will look very similar to those of the strong Kleene truth definition, but we have given up just enough of the compositionality to ensure closure under first order inference. The expanded format will allow us to register the connection between sub-sentences that are negations of each other. The result, as we shall show, is a truth definition where the strong Kleene truth definition has been strengthened with whatever is warranted by syntactic form. Before we give the formal definition we indicate how it differs from the strong Kleene truth definition.

In addition to atomic clauses corresponding to those of $\models$, we stipulate that $v \models\!\!= S^+, S^-$ is always the case, regardless of whether S is at all defined in v. In addition, the semantical clauses for disjunction allow us to infer $v \models\!\!= (\varphi \vee \psi)^+$ from $v \models\!\!= \varphi^+, \psi^+$. Hence we always have $v \models\!\!= (S \vee \neg S)^+$. In a similar way we will get $v \models\!\!= \varphi^+$ for any classical tautology φ.

Let 'Γ' range over finite sets of signed $\mathbf{l}[\rho_v]$ sentences φ^+ and φ^-. $\models\!\!=$ is the smallest relation satisfying the following:

$$v \models\!\!= \Gamma, \top^+.$$

$$v \models\!\!= \Gamma, S^+, S^- \quad \text{for all} \quad S \in \rho_v.$$

$$v \models\!\!= \Gamma, S^+ \quad \text{if} \quad S \in v^+.$$

$$v \models\!\!= \Gamma, S^- \quad \text{if} \quad S \in v^-.$$

$$v \models\!\!= \Gamma, \neg\varphi^+ \quad \text{if} \quad v \models\!\!= \Gamma, \varphi^-.$$

$$v \models\!\!= \Gamma, \neg\varphi^- \ \text{ if } \ v \models\!\!= \Gamma, \varphi^+.$$

$$v \models\!\!= \Gamma, (\varphi \vee \psi)^+ \ \text{ if } \ v \models\!\!= \Gamma, \varphi^+, \psi^+.$$

$$v \models\!\!= \Gamma, (\varphi \vee \psi)^- \ \text{ if } \ v \models\!\!= \Gamma, \varphi^- \ \text{ and } \ v \models\!\!= \Gamma, \psi^-.$$

Since the defining schema for $\models\!\!=$ is a generalization of the defining schema for $\models$, we shall use the label 'generalized strong Kleene truth definition.' But care is needed. A truth definition, as we shall use the term, applies to single signed sentences rather than finite sets of such. Let $\models\!\!=^*$ be the restriction of $\models\!\!=$ to single signed sentences. So $\models\!\!=^*$ is defined by the equivalence

$$v \models\!\!=^* \phi \quad \text{iff} \quad v \models\!\!= \phi$$

for all v and ϕ. By the *generalized strong Kleene truth definition* we shall mean $\models\!\!=^*$. Note especially that $\models\!\!=^*$ is *not* defined as the weakening of $\models\!\!=$ where the restriction to singletons is enforced throughout the induction. The following is a valid "verification tree:"

$$v \models\!\!= S^+, S^-$$

$$|$$

$$v \models\!\!= S^+, \neg S^+$$

$$|$$

$$v \models\!\!= (S \vee \neg S)^+$$

Hence $M \models\!\!=^* (S \vee \neg S)^+$ even if $M \not\models\!\!=^* S^+, S^-$ and $M \not\models\!\!=^* S^+, \neg S^+$.

In the remainder of this section we prove that $\models_\Box$, $\models\!\!=^*$ and $\models^{scl}$ coincide in propositional logic. This follows from the three results $\models_\Box \leq \models\!\!=^*$, $\models\!\!=^* \leq \models^{scl}$ and $\models^{scl} \leq \models_\Box$, which we now proceed to prove. But first a few definitions: For any φ, let c_φ be the *operator depth* of φ:

$$c_S = c_\top = 0$$

$$c_{(\varphi \vee \psi)} = max(c_\varphi, c_\psi) + 1$$

$$c_{\neg\varphi} = c_\varphi + 1$$

For a finite set Π of sentences we define σ_Π as $\Sigma_{\varphi \in \Pi} c_\varphi$, and for a finite set Γ of *signed* sentences we define σ_Γ as $\Sigma_{\varphi^+ \in \Gamma} c_\varphi + \Sigma_{\varphi^- \in \Gamma} c_\varphi$.

For finite sets $\Pi = \{\varphi_1, \ldots, \varphi_n\}$ of sentences, let $\bigvee\!\!\!\bigvee \Pi$ be the sentence $\varphi_1 \vee \ldots \vee \varphi_n$. $\bigvee\!\!\!\bigvee \emptyset = \bot$. For finite sets Γ of signed sentences, let $\bigvee\!\!\!\bigvee \Gamma$ be the sentence $\bigvee\!\!\!\bigvee(\{\varphi \mid \varphi^+ \in \Gamma\} \cup \{\neg\varphi \mid \varphi^- \in \Gamma\})$.

Theorem. $\models_\Box \leq \models\!\!=^*$.

This follows from the next, more general result.

Lemma. *If* $v \models_{\Box} \bigvee\!\!\!\!\bigvee\Gamma$, *then* $v \models\!\!= \Gamma$.

Proof: By induction on σ_Γ.

- Suppose $\sigma_\Gamma = 0$. Then Γ is a set of signed sentence symbols, and possibly containing $\top^+$ or $\top^-$. Moreover, suppose $v \not\models\!\!= \Gamma$. Then
  - $\top^+ \notin \Gamma$
  - $S^+ \in \Gamma \to S \notin v^+$
  - $S^- \in \Gamma \to S \notin v^-$
  - $\{S^+, S^-\} \subseteq \Gamma$ for no S.

  In this case, v also has a completion v' satisfying the same conditions. Consequently, $v' \not\models \bigvee\!\!\!\!\bigvee\Gamma$ and $v \not\models_{\Box} \bigvee\!\!\!\!\bigvee\Gamma$.

- Suppose $v \models_{\Box} \bigvee\!\!\!\!\bigvee(\Gamma_0, (\varphi_0 \vee \varphi_1)^+)$. Then clearly $v \models_{\Box} \bigvee\!\!\!\!\bigvee(\Gamma_0, \varphi_0^+, \varphi_1^+)$, and by the induction hypothesis $v \models\!\!= \Gamma_0, \varphi_0^+, \varphi_1^+$. Consequently $v \models\!\!= \Gamma_0, (\varphi_0 \vee \varphi_1)^+$.

- Suppose $v \models_{\Box} \bigvee\!\!\!\!\bigvee(\Gamma_0, (\varphi_0 \vee \varphi_1)^-)$. Then by familiar principles of classical logic it follows that $v \models_{\Box} \bigvee\!\!\!\!\bigvee(\Gamma_0, \varphi_0^-)$ and $v \models_{\Box} \bigvee\!\!\!\!\bigvee(\Gamma_0, \varphi_1^-)$. By the induction hypothesis $v \models\!\!= \Gamma_0, \varphi_0^-$ and $v \models\!\!= \Gamma_0, \varphi_1^-$. Hence $v \models\!\!= \Gamma_0, (\varphi_0 \vee \varphi_1)^-$.

- The induction steps for $\neg$ are left to the reader. □

The reader will note that this proof is a lot like a completeness proof for a deductive calculus relative to the set of classical tautologies. Using a slightly more complex induction, we could have imitated the method of semantic tableaux, starting with the assumption $v \not\models\!\!= \Gamma$ and building up a completion of v that constitutes a counterexample to $v \models_{\Box} \bigvee\!\!\!\!\bigvee\Gamma$. This is what we do when we turn to predicate logic and prove the relation $\models_{\Box\aleph_0} \leq \models\!\!=^*$. (The way we define $\models\!\!=$ for predicate logic, the relation $\models_{\Box} \leq \models\!\!=^*$ will no longer hold; $\models_{\Box\aleph_0}$ represents a sufficient weakening of $\models_{\Box}$.) We now proceed to the next link in our circular chain:

Theorem. $\models\!\!=^* \leq \models^{scl}$.

This is seen to follow from the next lemma, since negative equivalence yields classical equivalence, and thus classical implication.

Lemma. *If* $v \models\!\!= \Gamma$, *then* $v \models \chi$ *for a sentence* χ *such that* $\models_3 \neg\chi \equiv \neg\bigvee\!\!\!\!\bigvee\Gamma$.

Proof: By induction on $\models\!\!=$. We follow the clauses in the definition of $\models\!\!=$.

- If $\{S^+, S^-\} \subseteq \Gamma$ for some S, let χ be $\top$.
- If $v \models\!\!= \Gamma$ by one of the remaining atomic clauses, let χ be $\bigvee\!\!\!\!\bigvee \Gamma$ itself.
- In both the positive and negative step for negation, we can use the same χ given by the induction hypothesis.
- In the positive step for disjunction we can use the same χ given by the induction hypothesis.
- For the negative step for disjunction, suppose the last step is of the form

$$\frac{v \models\!\!= \Gamma_0, \varphi_0^- \qquad v \models\!\!= \Gamma_0, \varphi_1^-}{v \models\!\!= \Gamma_0, (\varphi_0 \vee \varphi_1)^-.}$$

  By the induction hypothesis there are sentences χ_0 and χ_1 such that $v \models (\chi_0 \wedge \chi_1)^+$, $\models_3 \neg\chi_0 \equiv \neg \bigvee\!\!\!\!\bigvee(\Gamma_0, \varphi_0^-)$ and $\models_3 \neg\chi_1 \equiv \neg \bigvee\!\!\!\!\bigvee(\Gamma_0, \varphi_1^-)$. Simple calculations now show that $\models_3 \neg(\chi_0 \wedge \chi_1) \equiv \neg \bigvee\!\!\!\!\bigvee(\Gamma_0, (\varphi_0 \vee \varphi_1)^-)$ □

The final link in our circular chain we get almost for free:

Theorem. $\models^{scl} \leq \models_\Box$.

Proof: This follows if $\models_\Box$ satisfies the closure conditions for $\models^{scl}$. By persistence, $v \models \phi$ implies $v \models_\Box \phi$. This takes care of the first condition. Since classical logic applies for complete models, $\models_\Box$ also satisfies the second condition. □

Hence we can sum up:

Theorem. $\models^{scl} = \models_\Box = \models\!\!=^*$.

1.7 Predictive and Saturated Sentences

A sentence φ is *predictive* if it satisfies the equivalence

$$v \models_\Box \varphi \quad \leftrightarrow \quad v \models \varphi.$$

So the class of predictive sentences is the class of sentences for which the truth definitions $\models_\Box$ and $\models$ coincide. The "size" of this class is a measure of the difference between $\models_\Box$ and $\models$. In this section we prove that every sentence is classically equivalent to a predictive sentence. We have assumed that similarity types ρ are finite, and this assumption comes in handy in the proof. It is not essential, however, as is seen from the following lemma, which holds for infinite as well as as finite ρ and ρ'. The lemma is a simple consequence of the corresponding fact about $\models$:

Lemma. *If v is a ρ' model, $\rho \subseteq \rho'$ and $\phi \in \mathbf{l}[\rho]$, then*

$$v \models_{\Box} \phi \quad \textit{iff} \quad v \upharpoonright \rho \models_{\Box} \phi.$$

Theorem. *Every $\mathbf{l}[\rho]$ sentence is negatively equivalent to a predictive $\mathbf{l}[\rho]$ sentence.*

Proof: For any $\mathbf{l}[\rho]$ sentence φ, let V_φ be the set

$$\{v \mid v \models_{\Box} \varphi\},$$

where 'v' ranges over ρ models. Since $\models_{\Box} = \models^*$, by the last lemma of the previous section there is for every $v \in V_\varphi$ an $\mathbf{l}[\rho]$ sentence χ_v such that $v \models \chi_v$ and $\models_3 \neg\varphi \equiv \neg\chi_v$. Since ρ is finite, so is V_φ, and hence $\chi = \bigvee_{v \in V_\varphi} \chi_v$ is a sentence. Clearly $\models_3 \neg\varphi \equiv \neg\chi$. To prove that χ is predictive, let u be an arbitrary ρ models such that $u \models_{\Box} \chi$. Then $u \models_{\Box} \varphi$, so $u \in V_\varphi$. Hence $u \models \chi_u$, and thus $u \models \chi$. □

We have defined a sentence χ to be saturated if for all φ it satisfies the equivalence

$$\varphi \models_2 \chi \quad \leftrightarrow \quad \varphi \models_3 \chi.$$

The two notions of predictive and saturated sentences are clearly connected:

Theorem. *Every predictive sentence is saturated.*

Proof: Suppose χ is predictive, and suppose $\varphi \models_2 \chi$. Let v be an arbitrary model such that $v \models \varphi$. Then $v \models_{\Box} \chi$. Since χ is predictive, also $v \models \chi$. Since v was arbitrary, we have shown $\varphi \models_3 \chi$. □

From these results the saturation theorem follows, from which the relative saturation theorem trivially follows, from which again the "cut-and-glue" theorem follows, which was the missing link in our persistence characterization proof.

We shall later see that predictiveness implies saturatedness also in predicate logic. The uniform saturation theorem does not hold, however, and hence neither does a corresponding result about predictiveness. Hence in the case of predicate logic we shall arrive at the "cut-and-glue" theorem from a slightly different route. As a final observation before we turn to predicate logic, we note that a sentence is saturated iff it is predictive. This does not hold in predicate logic.

Theorem. *Every saturated sentence is predictive.*

Proof: Suppose φ is saturated. φ is classically equivalent to some predictive sentence χ. Now suppose $v \models_{\Box} \varphi$. Then also $v \models_{\Box} \chi$ and hence $v \models \chi$. Now since $\chi \models_2 \varphi$ and φ is saturated, also $\chi \models_3 \varphi$, and hence $v \models \varphi$. Since v was arbitrary, φ must itself be predictive. □

2

The Strong Kleene Truth Definition

We now leave the simple language $\mathbf{l}$ of the previous chapter. Many of the insights gained in that chapter will be useful when in the rest of the book we ask the same general questions about a richer language $\mathcal{L}$, where a symbol $\exists$ for existential quantification has been added.

In chapter 3 we compare alternative truth definitions for $\mathcal{L}$, and in particular we study their relation to the strong Kleene truth definition. In chapter 4 we study extensions of the language $\mathcal{L}$ under the strong Kleene truth definition, in particular we determine in what sense this language is *maximal.*

In the present chapter we develop some basic model theory for $\mathcal{L}$ under the strong Kleene truth definition; thereby we shall lay much of the groundwork for the results that appear in the two later chapters. Indeed, most of the results included in this chapter will have important applications later in the book.

2.1 Basic Notions

The first order language $\mathcal{L}$ has propositional connectives $\vee$ and $\neg$, the quantifier $\exists$, a countable, infinite list of individual variables and a symbol $=$ for identity. A *similarity type* ρ is a finite set of *relation symbols* R, where each R has an associated *arity* $n_R \leq \omega$. Hence we shall not consider function symbols or constants, and neither shall we consider *infinite* similarity types. Relative to a given similarity type ρ, $R(y_1, \ldots, y_m)$ is an atomic formula iff $y_1, \ldots, y_m$ are individual variables, $R \in \rho$, and $n_R = m$. Similarly, $(x = y)$ is an atomic formula iff x and y are variables. The definitions of $\mathcal{L}[\rho]$ formulas, free variables, sentences, etc., are the standard ones. Note that we now use 'ρ' to range over similarity types for $\mathcal{L}$, rather than for $\mathbf{l}$. We follow up such an "efficient" use of notations below, and recycle most symbols and expressions from the previous chapter, to refer now to appropriate counterparts in predicate logic.

A *structure* or *model* M for similarity type ρ consists of a non-empty domain $|M|$, and for each relation symbol $R \in \rho$ an ordered pair $R^M = \langle R^{M^+}, R^{M^-} \rangle$ of disjoint sets of n_R-tuples from $|M|$. We say that N *informationally extends* M, $M \ll N$, if M and N have the same domain, and $R^{M^+} \subseteq R^{N^+}$, $R^{M^-} \subseteq R^{N^-}$ for all relation symbols R. A model M is *complete* iff for every relation symbol R, $R^{M^+} \cup R^{M^-} = |M|^{n_R}$. If $M \ll N$ and N is complete, then N is an *informational completion* (or just *completion*) of M.

Above, we did not give a specific format or representation for the models. For all that was said, a given model M could be a model for more than one similarity type. If M is a model for ρ, is M then a model for any $\rho' \supseteq \rho$ as well? We assume that this is not so; in M is represented *the* similarity type ρ_M for which M is a model.

We use capital letters from the beginning of the alphabet to range over *variable assignments*, functions from the set of variables into the domain of the model. We define the strong Kleene truth definition so that it applies to signed pairs of formulas and variable assignments. The definition has the following clauses:

$M \models R(y_1, \ldots, y_n)[A]^+$ iff $\langle A(y_1), \ldots, A(y_n) \rangle \in R^{M^+}$.

$M \models R(y_1, \ldots, y_n)[A]^-$ iff $\langle A(y_1), \ldots, A(y_n) \rangle \in R^{M^-}$.

$M \models (x = y)[A]^+$ iff $A(x) = A(y)$.

$M \models (x = y)[A]^-$ iff $A(x) \neq A(y)$.

$M \models \neg\varphi[A]^+$ iff $M \models \varphi[A]^-$.

$M \models \neg\varphi[A]^-$ iff $M \models \varphi[A]^+$.

$M \models (\varphi \vee \psi)[A]^+$ iff $M \models \varphi[A]^+$ or $M \models \psi[A]^+$.

$M \models (\varphi \vee \psi)[A]^-$ iff $M \models \varphi[A]^-$ and $M \models \psi[A]^-$.

$M \models \exists x\varphi[A]^+$ iff $M \models \varphi[A']^+$ for an A' differing with A at most at x.

$M \models \exists x\varphi[A]^-$ iff $M \models \varphi[A']^-$ for all A' differing with A at most at x.

Note that $\top$ is not primitive in this language; it can be defined as $\exists x(x = x)$. Similarly, we treat $\bot$, $\forall x\varphi$ and $(\varphi \wedge \psi)$ as abbreviations of $\neg\exists x(x = x)$, $\neg\exists x\neg\varphi$ and $\neg(\neg\varphi \vee \neg\psi)$.

As in the previous chapter, we sometimes simplify notations, and write '$M \models \varphi[A]$' for '$M \models \varphi[A]^+$'. Hence when the polarity sign is suppressed, it is assumed to be positive. We note that with the truth definition $\models$, $\mathcal{L}$ is coherent, determinable and persistent. The proofs are analogous to their counterparts in propositional logic, and are left to the reader:

Theorem.

(i) *For any model M, variable assignment A and formula φ, it is* not *the case that both $M \models \varphi[A]^+$ and $M \models \varphi[A]^-$.*

(ii) *If M is a complete model, then for any φ and A either $M \models \varphi[A]^+$ or $M \models \varphi[A]^-$.*

(iii) *If $M \ll N$, then $M \models \varphi[A]^\odot$ implies $N \models \varphi[A]^\odot$.*

Hence for any φ, A and complete model M we have $M \models \varphi[A]^+$ or $M \models \varphi[A]^-$, and not both. Given a similarity type ρ, let 'M_{cl}' range over classical models for ρ. M_{cl} identifies a domain $|M_{cl}|$, and assigns to every n-ary relation symbol R an extension $R^{M_{cl}} \subseteq |M_{cl}|^n$. Consider the relation $\bowtie$ between classical models M_{cl} and complete partial models M:

$$M_{cl} \bowtie M \quad \text{iff} \quad (|M_{cl}| = |M| \quad \text{and} \quad R^{M_{cl}} = R^{M^+})$$

Let $\models_{cl}$ be the standard truth definition for classical structures. We have:

Lemma. *If $M_{cl} \bowtie M$, then for every formula φ and variable assignment A we have $M_{cl} \models_{cl} \varphi[A]$ iff $M \models \varphi[A]$.*

The proof is exactly analogous to its counterpart in propositional logic. Since the relation $\bowtie$ is only defined for complete models M of the partial format, and *they* all satisfy $R^{M^-} = |M|^n - R^{M^+}$, this is a 1-1 relation. Given this correspondence, we shall in the sequel talk about the complete structures of the partial format as if they were classical structures, and take advantage of results from classical logic whenever the opportunity arises.

If $\rho \subseteq \rho'$ and M is a ρ' model, then we define the restriction $M \uparrow \rho$ of M to ρ as the ρ model with domain equal to the domain of M, and satisfying the equation

$$R^{M \uparrow \rho} = R^M$$

for all $R \in \rho$. As in propositional logic, only the interpretation of symbols occurring in a sentence are relevant to its truth or falsity. Here, and in the sequel, we use '$\odot$' to range over $+$ and $-$. The following can be proved by a straightforward induction on formulas:

Lemma. *If M is ρ' model, $\rho \subseteq \rho'$ and φ is an $\mathcal{L}[\rho]$ formula, then*

$$M \models \varphi[A]^\odot \quad \textit{iff} \quad M \uparrow \rho \models \varphi[A]^\odot.$$

We shall again define a translation procedure into classical, complete logic. Again we need a notion of *generalized* structures. These satisfy all the stipulations that apply for proper partial structures except the *coherence condition*, that $R^{M^+} \cap R^{M^-} = \emptyset$ for all relation symbols R. 'M' will be used to range over generalized as well as proper models, but only when explicitly indicated. Hence 'for all M' will ordinarily mean 'for all proper models M'.

Given a similarity type $\rho = \{R_1, \ldots, R_n\}$, let ρ^* be the similarity type $\{R_1^+, \ldots, R_n^+, R_1^-, \ldots, R_n^-\}$ where both R_i^+ and R_i^- are of the same arity as R_i. Now there is a 1-1 correspondence between generalized structures M for ρ and classical structures M^* for ρ^*, given by

$$|M| = |M^*| \quad R^{M^+} = R^{+^{M^*}} \quad R^{M^-} = R^{-^{M^*}}.$$

Note that $R^{+^{M^*}}$ and $R^{-^{M^*}}$ may overlap. Let $\mathcal{L}_{\forall,\wedge}$ be the language where $\forall$ and $\wedge$ have been added to $\mathcal{L}$ as primitives. For $\mathcal{L}[\rho]$ formulas φ we define the positive and negative translations φ^* and φ_* in $\mathcal{L}_{\forall,\wedge}[\rho^*]$. The two different uses of $=$ below should not be the cause of any confusion:

$$(x = y)^* = (x = y).$$

$$(x = y)_* = \neg(x = y).$$

$$R(x_1, \ldots, x_n)^* = R^+(x_1, \ldots, x_n).$$

$$R(x_1, \ldots, x_n)_* = R^-(x_1, \ldots, x_n).$$

$$(\neg\varphi)^* = \varphi_*$$

$$(\neg\varphi)_* = \varphi^*$$

$$(\varphi \vee \psi)^* = (\varphi^* \vee \psi^*).$$

$$(\varphi \vee \psi)_* = (\varphi_* \wedge \psi_*).$$

$$(\exists x\varphi)^* = \exists x\varphi^*.$$

$$(\exists x\varphi)_* = \forall x\varphi_*.$$

Lemma. *An $\mathcal{L}_{\forall,\wedge}[\rho^*]$ formula χ is φ^* for some $\mathcal{L}[\rho]$ formula φ iff χ contains the negation symbol only directly in front of atomic identity formulas.*

Proof: It is immediately seen that φ^* is of this form for any $\mathcal{L}[\rho]$ formula φ. On the other hand, for the $\mathcal{L}_{\forall,\wedge}[\rho^*]$ formulas of this form we can define an *inverse* to $*$.

$$(x = y)^{-*} = (x = y).$$

$(\neg(x = y))^{-*} = \neg(x = y)$.

$R^+(x_1, \ldots, x_n)^{-*} = R(x_1, \ldots, x_n)$.

$R^-(x_1, \ldots, x_n)^{-*} = \neg R(x_1, \ldots, x_n)$.

$(\varphi \vee \psi)^{-*} = (\varphi^{-*} \vee \psi^{-*})$.

$(\varphi \wedge \psi)^{-*} = \neg(\neg(\varphi^{-*}) \vee \neg(\psi^{-*}))$.

$(\exists x\varphi)^{-*} = \exists x\varphi^{-*}$.

$(\forall x\varphi)^{-*} = \neg\exists x\neg(\varphi^{-*})$.

A straightforward induction shows that $(\varphi^{-*})^*$ is identical to φ. □

Note, however, that $^{-*}$ is not an inverse in the strictest sense; the equality $(\varphi^*)^{-*} = \varphi$ does not hold in general.

As in propositional logic, $\models_2$ is the restriction of $\models$ to complete models, while $\models_3$ and $\models_4$ are the strong Kleene truth definition for proper and generalized structures, respectively. Let 'Φ' and 'Ψ' range over finite sets of signed formulas $\varphi^\odot$. $\Phi \models_3 \Psi$ holds iff for every M and A such that $M \models \varphi[A]^\odot$ for all $\varphi^\odot \in \Phi$, $M \models \varphi[A]^\odot$ for some $\varphi^\odot \in \Psi$. Similarly for $\Phi \models_2 \Psi$ and $\Phi \models_4 \Psi$; in the first we quantify over complete models, and in the second over generalized models. Hence $\Phi \models_4 \Psi$ implies $\Phi \models_3 \Psi$, which in turn implies $\Phi \models_2 \Psi$. Let $(\varphi^+)^*$ and $(\varphi^-)^*$ be φ^* and φ_*. Finally, we define Φ^* as $\{\phi^* \mid \phi \in \Phi\}$.

Lemma. *For any generalized model M, formula φ, and sets Φ and Ψ of signed formulas*

(i) $M \models \varphi[A]^+$ *iff* $M^* \models \varphi^*[A]^+$.

(ii) $M \models \varphi[A]^-$ *iff* $M^* \models \varphi_*[A]^+$.

(iii) $\Phi \models_4 \Psi$ *iff* $\Phi^* \models_2 \Psi^*$.

Proof: (iii) follows from (i) and the 1-1 correspondence between classical and generalized structures. (i) and (ii) can be proved by a simultaneous induction on φ, and is left to the reader. □

The clauses for $\exists$ in the definition of $\models$ could be stated more simply with the use of some additional notation. When A is a variable assignment, a is an element from the corresponding model and x is a variable, then $A(a/x)$ is the variable assignment B, identical to A except that $B(x) = a$. Hence $M \models \exists x\varphi[A]^+$ iff $M \models \varphi[A(a/x)]^+$ for some $a \in |M|$. Analogously for $M \models \exists x\varphi[A]^-$.

As in classical logic, the values of A are irrelevant for variables not occurring free in φ. Given the translation procedure, this result follows trivially from its counterpart in classical logic:

Lemma. *If* $A(x) = B(x)$ *for all variables* x *occurring free in* φ, *then* $M \models \varphi[A]^{\odot}$ *iff* $M \models \varphi[B]^{\odot}$.

Let $\varphi(x/y)$ be the result of substituting x for all free occurrences of the variable y in φ. The next lemma resembles the previous, and is similarly a trivial corollary to its classical counterpart:

Lemma. *If* $A(x) = A(y)$, *and* x *does not occur in* φ, *then*

$$M \models \varphi(x/y)[A]^{\odot} \quad \textit{iff} \quad M \models \varphi[A]^{\odot}.$$

Let the n-tuple $x_1, \ldots, x_n$ of variables be implicitly given, and let the free variables of φ be among $x_1, \ldots, x_n$; in keeping with standard practice we shall write $M \models \varphi[a_1, \ldots, a_n]$ to indicate the existence of an A such that $M \models \varphi[A]$, where $A(x_i) = a_i$ for all $i : 1 \leq i \leq n$. If φ is a sentence, and hence independent of the A altogether, we may drop the '$[A]$' and write simply $M \models \varphi$.

We introduce a substitution function $\varphi(\psi, \chi/\alpha)$ for atomic formulas α, similarly to the one we used for **l**.

Definition.

$\alpha(\psi, \chi/\alpha) = \psi$.

$\beta(\psi, \chi/\alpha) = \beta$ *for atomic formulas* β *distinct from* α.

$(\neg\varphi)(\psi, \chi/\alpha) = \neg(\varphi(\chi, \psi/\alpha))$.

$(\varphi_0 \vee \varphi_1)(\psi, \chi/\alpha) = (\varphi_0(\psi, \chi/\alpha) \vee \varphi_1(\psi, \chi/\alpha))$.

$(\exists x\varphi)(\psi, \chi/\alpha) = \exists x(\varphi(\psi, \chi/\alpha))$.

Sentence symbols S are 0-ary relation symbols. We use this substitution function to obtain the strong and positive equivalence theorems from their classical counterpart, exactly as we did in propositional logic. We let $\varphi(\begin{smallmatrix}\psi & \chi \\ S & T\end{smallmatrix})$ be the result of a uniform, simultaneous substitution of ψ for S and χ for T in φ. Exactly as in chapter 1, we obtain:

Lemma. $(\varphi(\psi, \chi/S))^* = \varphi^*(\begin{smallmatrix}\psi^* & \chi_* \\ S^+ & S^-\end{smallmatrix})$ *and* $(\varphi(\psi, \chi/S))_* = \varphi_*(\begin{smallmatrix}\chi^* & \psi_* \\ S^+ & S^-\end{smallmatrix})$

The proof follows a straightforward induction, and is left to the reader. We shall use the following well known theorem from classical logic:

Theorem. *Suppose* S *and* T *only have positive occurrences in* φ. *Then for any complete model* M, *if*

$$M \models \psi_0 \supset \psi_1 \quad \textit{and} \quad M \models \chi_0 \supset \chi_1,$$

then

$$M \models \varphi(\begin{smallmatrix}\psi_0 & \chi_0 \\ S & T\end{smallmatrix}) \supset \varphi(\begin{smallmatrix}\psi_1 & \chi_1 \\ S & T\end{smallmatrix}).$$

As a counterpart, we now get the following *substitution theorem*, which holds for all $\mathcal{L}$ formulas φ.

Theorem.

(i) *If* $M \models \psi_0 \supset \psi_1$ *and* $M \models \neg\chi_0 \supset \neg\chi_1$, *then* $M \models \varphi(\psi_0, \chi_0/S) \supset \varphi(\psi_1, \chi_1/S)$.

(ii) *If* $\models_3 \psi_0 \supset \psi_1$ *and* $\models_3 \neg\chi_0 \supset \neg\chi_1$, *then* $\models_3 \varphi(\psi_0, \chi_0/S) \supset \varphi(\psi_1, \chi_1/S)$.

Proof: (ii) is an immediate consequence of (i), while (i) follows from the properties of * and the lemma and theorem above, as it did in propositional logic. We use the fact that for sentence symbols S, S^+ and S^- have only positive occurrences in φ^*. □

This also gives the positive and strong equivalence theorems as corollaries. For details, see chapter 1. As before, $\varphi(\psi/^+S)$ and $\varphi(\psi/S)$ are defined as $\varphi(\psi, S/S)$ and $\varphi(\psi, \psi/S)$.

Corollary. *If* $\models_3 \psi \equiv \chi$, *then* $\models_3 \varphi(\psi/^+S) \equiv \varphi(\chi/^+S)$.

Corollary. *If* $\models_3 \psi \rightleftharpoons \chi$, *then* $\models_3 \varphi(\psi/S) \rightleftharpoons \varphi(\chi/S)$.

2.2 Interpolation

In this section we use the substitution theorem to prove a counterpart to the Craig interpolation theorem. The proof is surprisingly simple, considering the fact that in classical first order logic the interpolation theorem is a much deeper result than this substitution theorem. But the two interpolation theorems are not quite comparable. Unlike in classical logic, both truth and falsity are primitive in the partial logic. Hence a consequence relation corresponding to

$$\varphi^+ \models_3 \psi^+ \quad \text{and} \quad \psi^- \models_3 \varphi^-$$

is perhaps a more worthy counterpart to the classical consequence relation. This is the line taken in Blamey (1980, 1986). The interpolation theorem for such a consequence relation does not seem to be as easily obtained as the interpolation theorem for $\models_3$. Here we are only interested in a particular application, however, and the present interpolation theorem is the one we need.

From a certain point of view the consequence relation $\models_3$ gives us a more transparent logic than does $\models_2$. This is not to say that $\models_3$ is *simpler*, since $\varphi \models_2 \psi$ can be coded as *compl*, $\varphi \models_3 \psi$, where *compl* is the sentence

$$\bigwedge_{R \in \rho} \forall x_1 \ldots \forall x_{n_R}(R(x_1, \ldots, x_{n_R}) \vee \neg R(x_1, \ldots, x_{n_R})).$$

But the consequence relation $\models_3$ carries interpolation on its sleeve in a way that $\models_2$ does not. If $\varphi \models_3 \psi$ and the relation symbol R, distinct from identity, does not occur in φ, then an atomic subformula $R(x_1, \ldots, x_n)$ of ψ does in no way contribute to ψ being a consequence of φ; it could as well be replaced by *anything*, and the consequence relation would still hold. To be stressed here is the fact that each individual occurrence of $R(x_1, \ldots, x_n)$ could be replaced by something else, independently of the other occurrences. In particular, the positive and the negative occurrences can be replaced by different formulas; we have the *independent substitution lemma*:

Lemma. *If Q is a relation symbol other than identity, not occurring in φ, and $\models_3 \varphi \supset \psi$, then for any $x_1, \ldots, x_n$ and any formulas χ_0 and χ_1 we have $\models_3 \varphi \supset (\psi(\chi_0, \chi_1/Q(x_1, \ldots, x_n)))$.*

Proof: Assume the antecedent of the lemma, and suppose $M \models \varphi[A]$. Now let M_φ be the model with the same domain as M, and with the same interpretation for relation symbols occurring in φ, but such that $R^{M_\varphi^+} = R^{M_\varphi^-} = \emptyset$ for relation symbols R *not* occurring in φ. Since $M \models \varphi[A]$, also $M_\varphi \models \varphi[A]$. We have assumed $\models_3 \varphi \supset \psi$, hence $M_\varphi \models \psi[A]$. Since Q does not occur in φ, $M_\varphi \not\models Q(x_1, \ldots, x_n)[B]^+$ and $M_\varphi \not\models Q(x_1, \ldots, x_n)[B]^-$ for all B. Hence $M_\varphi \models Q(x_1, \ldots, x_n) \supset \chi_0$ and $M_\varphi \models \neg Q(x_1, \ldots, x_n) \supset \neg\chi_1$. By the substitution theorem $M_\varphi \models \psi \supset (\psi(\chi_0, \chi_1/Q(x_1, \ldots, x_n)))$. Consequently $M_\varphi \models \psi(\chi_0, \chi_1/Q(x_1, \ldots, x_n))[A]$, and since $M_\varphi \ll M$, by persistence also $M \models \psi(\chi_0, \chi_1/Q(x_1, \ldots, x_n))[A]$. Since M and A were arbitrary, we have proved $\models_3 \varphi \supset (\psi(\chi_0, \chi_1/Q(x_1, \ldots, x_n)))$. □

We have defined $\top$ and $\bot$ as $\exists x(x = x)$ and $\neg\exists x(x = x)$ respectively; we observe that $M \models \top[A]^+$ and $M \models \bot[A]^-$ for all M and A. Hence also $\models_3 \bot \supset \alpha$ and $\models_3 \neg\top \supset \neg\alpha$ for any α. Now since $\psi(\alpha, \alpha/\alpha)$ is ψ itself, by an application of the substitution theorem we immediately obtain:

Lemma. *For any ψ and α, $\models_3 \psi(\bot, \top/\alpha) \supset \psi$.*

This last substitution is of particular interest. We define the *trivialization* of ψ over α, $\psi \triangledown \alpha$, as $\psi(\bot, \top/\alpha)$. Assuming an enumeration of atomic formulas, we extend this definition to finite sets Ω of atomic formulas other than identity formulas. Let $\sigma(\Omega)$ be the first element of Ω:

$$\psi \triangledown \emptyset = \psi$$

$$\psi \triangledown \Omega = (\psi \triangledown (\Omega - \{\sigma(\Omega)\})) \triangledown \sigma(\Omega) \qquad \text{if } \Omega \text{ is non-empty.}$$

It is convenient to define the trivialization for yet another category; if ρ is a finite set of relation symbols not including identity, let $\psi \triangledown \rho$ be $\psi \triangledown \Omega_\psi^\rho$, where Ω^{ρ}_{ψ} is the set of atomic formulas $R(x_1, \ldots, x_n)$ occurring in ψ, for which $R \in \rho$. We now state the *interpolation theorem.* (ρ_φ is the set of relation symbols other than identity, with occurrences in φ.)

Theorem. *If* $\models_3 \varphi \supset \psi$, *then* $\models_3 \varphi \supset \chi$ *and* $\models_3 \chi \supset \psi$ *for some* $\mathcal{L}[\rho_\varphi \cap \rho_\psi]$ *formula* χ.

Proof: Simple arguments show that $\psi \nabla (\rho_\psi - \rho_\varphi)$ is an $\mathcal{L}[\rho_\varphi]$ formula as well as an $\mathcal{L}[\rho_\psi]$ formula, and hence an $\mathcal{L}[\rho_\varphi \cap \rho_\psi]$ formula. Moreover, by repeated applications of the lemmas above, we get

$$\models_3 \psi \nabla (\rho_\psi - \rho_\varphi) \supset \psi$$

and

$$\models_3 \varphi \supset \psi \quad \rightarrow \quad \models_3 \varphi \supset \psi \nabla (\rho_\psi - \rho_\varphi).$$

Hence $\psi \nabla (\rho_\psi - \rho_\varphi)$ is an interpolant. □

Note that we actually give an effective procedure to find the interpolant $interp(\varphi, \psi) = \psi \nabla (\rho_\psi - \rho_\varphi)$. Moreover, $interp(\varphi, \psi)$ is defined regardless of whether $\models_3 \varphi \supset \psi$. Note also that $interp(\top, \psi) \in \mathcal{L}[\emptyset]$, and for every ψ

$$\models_3 interp(\top, \psi) \supset \psi$$

and

$$\models_3 \top \supset \psi \quad \rightarrow \quad \models_3 \top \supset interp(\top, \psi).$$

Hence $\models_3 \psi$ iff $\models_3 interp(\top, \psi)$, and we have *almost* proved the following theorem:

Theorem. *The set* $\{\varphi \mid \models_3 \varphi\}$ *of valid formulas with respect to* $\models_3$ *is recursive.*

Proof: In view of the above remarks, this follows from the next lemma, together with the fact that the set of classically valid pure identity formulas has a recursive characteristic function.

Lemma. *If* φ *is an* $\mathcal{L}[\emptyset]$ *formula, then* $\models_3 \varphi$ *iff* $\models_2 \varphi$.

Proof: Let φ be an $\mathcal{L}[\emptyset]$ formula. Clearly, if $\models_3 \varphi$ then $\models_2 \varphi$. To prove the other direction, suppose $\not\models_3 \varphi$, i.e., $M \not\models \varphi[A]$ for some M and A. Then $M \uparrow \emptyset \not\models \varphi[A]$. $M \uparrow \emptyset$ is an $\emptyset$ model, hence vacuously it is complete. Hence $\not\models_2 \varphi$. □

It is well known that the analogue of the last theorem fails for classical logic, and it should be obvious where an analogous argument would break down.

The heart of the interpolation argument is the independent substitution lemma, the proof of which makes essential use of the partiality. To see that the analogue of this lemma is false of classical logic, note that for any φ

$$\models_2 \varphi \supset (R(x) \vee \neg R(x)),$$

even for R that do not occur in φ. Hence *if* an analogue of the independent substitution lemma were to hold, we would also get

$$\models_2 \varphi \supset (R(x) \vee \neg R(x))(\bot, \top / R(x)),$$

i.e.,

$$\models_2 \varphi \supset (\bot \vee \neg \top),$$

which of course does not hold for formulas φ in general.

Note also that the general *consequence relation* $\models_3$ is not recursive, i.e., the set $\{\langle \varphi, \psi \rangle \mid \varphi \models_3 \psi\}$ of pairs of $\mathcal{L}[\rho]$ formulas is not recursive for ρ in general. To see this, observe that $\models_2 \varphi$ iff *compl* $\models_3 \varphi$, where *compl* is the sentence

$$\bigwedge_{R \in \rho} \forall x_1 \ldots \forall x_n (R(x_1, \ldots, x_{n_R}) \vee \neg R(x_1, \ldots, x_{n_R})).$$

Since validity is recursive and the consequence relation is not, there cannot be anything resembling a *deduction theorem* in this system; in particular $\varphi \models_3 \psi$ is *not* equivalent to $\models_3 \neg\varphi \vee \psi$. New propositional connectives can buy us a deduction theorem; in particular we have already observed that $\varphi \models_3 \psi$ iff $\models_3 \varphi \supset \psi$.

2.3 Domain Persistence

In the next chapter we shall need a characterization of the sentences that are preserved under extension of domain. To discuss this, we need some new terminology. We have previously defined the restriction $M \uparrow \rho$ of a model M to a smaller similarity type ρ. We now introduce another type of restriction.

Definition. *For any non-empty subset D of $|M|$, let the* restriction $M \uparrow D$ *be the structure with domain D, satisfying the equations*

$$R^{M \uparrow D^{\odot}} = R^{M^{\odot}} \cap D^{n_R}.$$

M' is a substructure of M iff $M' = M \uparrow D$ for some non-empty $D \subseteq |M|$.

Definition. *A formula φ is (positively)* preserved under extension of domain *iff for every structure M, finite subset D of $|M|$ and variable assignment A into D, if $M \uparrow D \models \varphi[A]^+$, then $M \models \varphi[A]^+$. The corresponding definitions for classical and generalized models are similar.*

An *existential* sentence is a sentence of the form $\exists x_1 \ldots \exists x_n \varphi$, where $n \geq 0$ and φ is quantifier free. A well known characterization result of classical logic states that a sentence is preserved under extension of domain for classical structures iff it is classically equivalent to an existential sentence. In this section we prove a corresponding result about partial structures. As a step towards this theorem, we also prove the analogous characterization result for generalized structures. The following correspondence result will be useful:

Lemma. *A formula φ is preserved under extension of domain for generalized partial structures iff φ^* is preserved under extension of domain for classical structures.*

Proof: Since $(M \uparrow D)^* = M^* \uparrow D$, this follows immediately from what we know about the translation function. □

Definition. *A formula φ is* increasing *if for every two classical models M and N and variable assignment A, if $|M| = |N|$ and $R^M \subseteq R^N$ for every R, then $M \models \varphi[A]$ implies $N \models \varphi[A]$.*

We use the following results from classical logic. The first is well known, and follows by a straightforward induction.

Lemma. *An $\mathcal{L}_{\forall,\wedge}$ formula φ is increasing if the negation symbol only occurs directly in front of atomic identity formulas in φ.*

Theorem. *An $\mathcal{L}_{\forall,\wedge}$ sentence is increasing and preserved under extension of domain for classical models iff it is classically equivalent to an existential sentence in which the negation symbol only occurs directly in front of atomic identity formulas.*

Proof: One direction is well known. Half of this direction is just the above lemma, as restricted to sentences. And by a straightforward induction, it also follows that an existential formula is preserved under extension of domain.

We shall prove the other direction from a particular interpolation theorem, using a technique due to R. Lyndon. In exercise 2.2.19 of Chang and Keisler (1973) we find the following variant of the Craig–Lyndon interpolation theorem. ($Rel^+(\varphi)$ and $Rel^-(\varphi)$ are the sets of relation symbols with positive and negative occurrences, respectively, in φ.)

Theorem. *Let φ and ψ be $\mathcal{L}_{\forall,\wedge}$ sentences, and let ψ be existential. If $\varphi \models_2 \psi$, then there is an existential sentence χ such that $\varphi \models_2 \chi$, $\chi \models_2 \psi$, $Rel^+(\chi) \subseteq Rel^+(\varphi) \cap Rel^+(\psi)$ and $Rel^-(\chi) \subseteq Rel^-(\varphi) \cap Rel^-(\psi)$.*

From this interpolation theorem we shall derive the characterization of sentences that are both increasing and has the additional property of being preserved under extension of domain, using exactly the argument used in Lyndon (1959b) to characterize increasing formulas, in that case starting with a correspondingly less specialized interpolation theorem.

So suppose the sentence φ is increasing and preserved under extension of domain. By a well known characterization theorem, φ is classically equivalent to an existential sentence ψ. Now ψ is also increasing. Hence

$$\psi \wedge I \models_2 \psi',$$

where I is the sentence

$$\bigwedge_R \forall x_1 \ldots \forall x_n (R(x_1, \ldots, x_{n_R}) \rightarrow R'(x_1, \ldots, x_{n_R})).$$

Here, R ranges over relation symbols of the similarity type and R' is a new relation symbol, unique for each R. As before, '$\alpha \rightarrow \beta$' abbreviates '$\neg\alpha \vee \beta$'. ψ' is ψ with each R replaced by R'. Clearly ψ' is an existential sentence. By the above interpolation theorem there is an existential sentence χ such that

$$\psi \wedge I \models_2 \chi \qquad \chi \models_2 \psi' \qquad Rel^-(\chi) \subseteq Rel^-(I \wedge \psi) \cap Rel^-(\psi').$$

Clearly $Rel^-(I \wedge \psi) \cap Rel^-(\psi') = \emptyset$, hence at most the identity symbol has negative occurrences in χ. We can assume that the negation symbol only occurs directly in front of atomic identity formulas in χ.

Replacing the primed relation symbols with their unprimed counterparts, we obtain from χ a new existential sentence χ_1, also in which the negation symbol only occurs directly in front of atomic identity formulas. Moreover, by this substitution we obtain ψ from ψ', ψ remains unchanged, while I turns into a theorem of classical predicated logic. Since we had

$$\psi \wedge I \models_2 \chi \quad \text{and} \quad \chi \models_2 \psi',$$

it follows that

$$\psi \models_2 \chi_1 \quad \text{and} \quad \chi_1 \models_2 \psi,$$

or, equivalently,

$$\models_2 \chi_1 \equiv \psi, \quad \text{and hence} \quad \models_2 \chi_1 \equiv \varphi. \qquad \square$$

Since the map * deletes occurrences of double negation, it is not always the case that φ is an existential sentence iff φ^* is. However, we *do* have the following:

Lemma. *If φ is an existential sentence, then so is φ^*. If φ is an existential sentence and φ^{-*} is defined, then φ^{-*} is existential.*

Proof: This is read directly off the definitions of * and $^{-*}$. □

Lemma. *An $\mathcal{L}$ sentence φ is preserved under extension of domain for generalized structures iff there exists an existential $\mathcal{L}$ sentence ψ such that $\models_4 \varphi \equiv \psi$.*

Proof: Equivalence is already established between consecutive items below:

1. φ is preserved under extension of domain for generalized structures.
2. φ^* is preserved under extension of domain for classical structures.
3. φ^* is both increasing and preserved under extension of domain for classical structures.
4. There is an existential $\mathcal{L}_{\forall,\wedge}[\rho^*]$ sentence χ with the negation symbol only occurring directly in front of atomic identity formulas, such that $\models_2 \varphi^* \equiv \chi$.
5. There is an existential $\mathcal{L}[\rho]$ sentence ψ such that $\models_2 \varphi^* \equiv \psi^*$.
6. There is an existential $\mathcal{L}[\rho]$ sentence ψ such that $\models_4 \varphi \equiv \psi$. □

Having proved this result for generalized structures, we proceed towards a corresponding result for proper structures.

Definition. *For any given (finite) similarity type ρ, let contr be the sentence*

$$\bigvee_{R\in\rho} \exists x_1 \ldots \exists x_{n_R}(R(x_1,\ldots,x_{n_R}) \wedge \neg R(x_1,\ldots,x_{n_R})).$$

Lemma. *φ is preserved under extension of domain for partial structures iff $(\varphi \vee contr)$ is preserved under extension of domain for generalized structures.*

Proof:

- First suppose φ is preserved under extension of domain for proper partial structures, and suppose $M \uparrow D \models (\varphi \vee contr)[A]^+$ for the generalized structure M.
  - If M is not a proper structure, then $M \models contr$, so $M \models (\varphi \vee contr)[A]^+$.
  - If M *is* a proper structure, then so is $M \uparrow D$. Hence $M \uparrow D \models \varphi[A]^+$, so by assumption, $M \models \varphi[A]^+$. Hence $M \models (\varphi \vee contr)[A]^+$.

- Next suppose $(\varphi \vee contr)$ is preserved under extension of domain for generalized structures. Then trivially $(\varphi \vee contr)$ is also preserved under extension of domain for proper structures. Hence so is φ, since $\models_3 \varphi \equiv (\varphi \vee contr)$. □

Lemma. $\models_3 \varphi \equiv \psi$ *iff* $\models_4 (\varphi \vee contr) \equiv (\psi \vee contr)$.

Proof: This is trivial, since $M \models contr$ iff M is not a proper partial structure. □

Theorem. *An $\mathcal{L}$ sentence φ is preserved under extension of domain for partial structures iff there is an existential $\mathcal{L}$ sentence ψ such that $\models_3 \varphi \equiv \psi$.*

Proof:

- First suppose φ is preserved under extension of domain for proper partial structures. Then $(\varphi \vee contr)$ is preserved under extension of domain for generalized structures. Hence $\models_4 (\varphi \vee contr) \equiv \chi$ for an existential sentence χ. But then clearly $\models_3 \varphi \equiv \chi$.

- Next suppose $\models_3 \varphi \equiv \chi$ for an existential sentence χ. Then also $\models_4 (\varphi \vee contr) \equiv (\chi \vee contr)$. $(\chi \vee contr)$ is a disjunction of existential sentences, hence it is a disjunction of sentences preserved under extension of domain for generalized structures. But then $(\chi \vee contr)$ is clearly itself preserved under extension of domain for generalized structures. Hence so is $(\varphi \vee contr)$, so φ is preserved under extension of domain for *proper* partial structures. □

The property characterized above is of interest in a general study of persistence. In this book we assume a rather special framework where the domain of individuals is assumed to be well known and understood, and where the introduction of new individuals represents a contradiction of previous information. As a result, we defined $M \ll N$ to imply $|M| = |N|$. An alternative interpretation of the framework would motivate a weaker notion where identity of domains is replaced by inclusion:

$$M \sqsubseteq N \quad \text{iff} \quad (|M| \subseteq |N| \text{ and } R^{M^{\odot}} \subseteq R^{N^{\odot}} \text{ for all } R \in \rho).$$

Definition. *A formula φ is domain truth-persistent iff $M \models \varphi[A]^+$ and $M \sqsubseteq N$ together imply $N \models \varphi[A]^+$ for all M, N, A. Domain falsity-persistence is defined analogously. A formula is domain persistent iff it is both domain truth-persistent and domain falsity-persistent.*

Clearly

$$M \ll N \quad \text{implies} \quad M \sqsubseteq N$$

and

$$M = N \uparrow D \quad \text{implies } M \sqsubseteq N,$$

while

$$M \sqsubseteq N \quad \text{implies} \quad (M \ll O \text{ and } O = N \uparrow D \text{ for some } O \text{ and } D).$$

Hence a formula is domain truth-persistent iff it is both truth-persistent and preserved under extension of domain.

Above we only considered the preservation of *truth* under extension of domain. We say that a formula is *strongly* preserved under extension of domain iff both truth and falsity is preserved in this way. Hence a formula is domain persistent iff it is both persistent and strongly preserved under extension of domain.

In chapter 4 we show that the coherent, determinable and persistent sentences of various languages are all strongly equivalent to $\mathcal{L}$ sentences. Hence coherence, determinability and *domain* persistence will be characterized by a *subset* of the $\mathcal{L}$ sentences. We shall leave it as an open problem to identify such a characterization, but we make a few observations.

We know that the domain truth-persistent sentences of $\mathcal{L}$ are positively equivalent to existential $\mathcal{L}$ sentences. An easy corollary states that the domain falsity-persistent sentences are negatively equivalent to universal $\mathcal{L}$ sentences. Hence a domain persistent sentence will be positively equivalent to an existential sentence, and negatively equivalent to a universal sentence. Can we improve on this result? One might conjecture that a domain persistent sentence is strongly equivalent to a quantifier free formula, but this is not the case, as is seen from the next example.

Example. *Consider the sentence $\varphi = \exists x(P(x) \vee \neg P(x))$. Clearly this sentence is positively preserved under extension of domain. Moreover, it is never false, and thus vacuously it is also negatively preserved and hence strongly preserved under extension of domain. We show that the sentence is not strongly (or even positively) equivalent to any quantifier free formula. Let M be a model with two elements, $|M| = \{0,1\}$. $P^{M^+} = \{0\}$, $P^{M^-} = \emptyset$. A is the constant function to 1. We have $M \models \varphi[A]^+$. Now suppose $\models_3 \varphi \equiv \psi$ for some quantifier free formula ψ. Then $M \models \psi[A]^+$. Now $A(y) \neq 0$ for all variables y. Since ψ is quantifier free, it is easy to see that $N \models \psi[A]^+$ for the model N differing from M only in that $P^{N^+} = \emptyset$. But this is impossible, since $N \not\models \varphi[A]^+$, and φ and ψ were assumed to be positively equivalent.*

2.4 Compactness

Finally we show a compactness result for $\models_3$. Since

$$\Phi \models_3 \Psi, \varphi^+$$

is not equivalent to

$$\Phi, \varphi^- \models_3 \Psi$$

in general, the following symmetrical result is *not* a trivial consequence of either of its one sided counterparts. In a later chapter we shall study a language that is compact in a classical or "one-sided" sense, but not in this stronger sense.

Theorem. *If* $\Phi \models_3 \Psi$ *then* $\Phi_0 \models_3 \Psi_0$ *for some finite subsets* Φ_0 *and* Ψ_0 *of* Φ *and* Ψ.

Proof: A generalized model M is proper iff $M \not\models$ *contr*. Therefore, since

$$\Phi \models_3 \Psi$$

also

$$\Phi \models_4 \Psi, contr.$$

By the behavior of *

$$\Phi^* \models_2 \Psi^*, contr^*.$$

By compactness of standard first order logic

$$(\Phi_0)^* \models_2 (\Psi_0)^*, contr^*$$

for some finite subsets Φ_0 and Ψ_0 of Φ and Ψ. Hence

$$\Phi_0 \models_4 \Psi_0, contr.$$

By the properties of *contr*, also

$$\Phi_0 \models_3 \Psi_0.$$

□

2.5 Semantical Equivalence for Structures

One of the major concerns in model theory is to gauge the expressive power of a given formal language; to determine what features of a model are "describable" with sentences from that formal language. In the present context it is of interest to study how the expressive power of $\mathcal{L}$ changes as we pass between some of the alternative truth definitions.

In the next chapter we present some evidence that such a change really does take place, while in the present section we try to determine the expressive power of $\mathcal{L}$ under $\models$ through an algebraic characterization of *semantical equivalence* between structures, to be defined below. We start with some useful algebraic notions:

In classical logic, two structures (for a similarity type without constants and function symbols) are *isomorphic*, written $M \cong N$, if there exists a bijection g from $|M|$ onto $|N|$ such that for all R and all n_R-tuples $\langle a_1, \ldots, a_{n_R}\rangle$ from $|M|$

$$\langle a_1, \ldots, a_{n_R}\rangle \in R^M \quad \text{iff} \quad \langle g(a_1), \ldots, g(a_{n_R})\rangle \in R^N.$$

There are at least two ways to generalize this to partial structures, one stronger than the other. We reserve the name "isomorphism" for the strongest of these:

Definition. *The function g is an* isomorphism *from the partial structure M to the partial structure N, $g : M \cong N$, iff g is a bijection from $|M|$ onto $|N|$ and the following holds for all relation symbols R and all n_R-tuples $\langle a_1, \ldots, a_{n_R}\rangle$ from $|M|$:*

$$\langle a_1, \ldots, a_{n_R}\rangle \in R^{M^{\odot}} \quad \textit{iff} \quad \langle g(a_1), \ldots, g(a_{n_R})\rangle \in R^{N^{\odot}}.$$

M and N are isomorphic, $M \cong N$, iff $g : M \cong N$ for some g.

The second generalization is obtained by substituting 'only if' for 'iff' in the above. Instead of using a more proper but cumbersome name like 'monotonic bijection' we choose the name 'monomorphism':

Definition. *The function g is a* monomorphism *from the partial structure M to the partial structure N, $g : M \widetilde{\ll} N$, iff g is a bijection from $|M|$ onto $|N|$ and the following holds for all relation symbols R and n_R-tuples $\langle a_1, .., a_{n_R}\rangle$ from $|M|$:*

$$\textit{If } \langle a_1, \ldots, a_{n_R}\rangle \in R^{M^{\odot}}, \quad \textit{then} \quad \langle g(a_1), \ldots, g(a_{n_R})\rangle \in R^{N^{\odot}}.$$

M is monomorphic to *N, $M \widetilde{\ll} N$, iff $g : M \widetilde{\ll} N$ for some g.*

It is easy to see that $M \widetilde{\ll} N$ iff there exists a structure M' such that $M \ll M'$ and $M' \cong N$; this motivates the symbol '$\widetilde{\ll}$'. By '$dom(p)$' and '$rng(p)$' we mean the domain and range of p.

Definition. *Let $I = \langle I_n\rangle_{0 \le n < \omega}$ be a sequence of non-empty sets of monomorphisms from substructures of M to substructures of N. I is an ω-*partial monomorphism *from M to N, $I : M \widetilde{\ll}_\omega N$, iff I satisfies the following* back- and forth- *properties:*

- *For any $q \in I_{n+1}$ and $a \in |M|$, there is a $p \in I_n$ such that $q \subseteq p$ and $a \in dom(p)$.*
- *For any $q \in I_{n+1}$ and $b \in |N|$, there is a $p \in I_n$ such that $q \subseteq p$ and $b \in rng(p)$.*

M is ω-partially monomorphic to N, $M \widetilde{\ll}_\omega N$, iff $I : M \widetilde{\ll}_\omega N$ for some I.

Substituting 'isomorphism' for 'monomorphism' in the above, we obtain the notion of *ω-partial isomorphisms*, written $\cong_\omega$.

Having developed these algebraic notions, we go on to define some semantic ones:

Definition. *N is* semantically at least as strong as *M, $M \overline{\ll} N$, if $M \models \phi$ implies $N \models \phi$ for all signed $\mathcal{L}$ sentences ϕ.*

Definition. *M and N are* semantically equivalent, *$M \equiv N$, if $M \overline{\ll} N$ and $N \overline{\ll} M$.*

We shall prove that N is semantically at least as strong as M iff there exists an ω-partial monomorphism from M to N. The argument is by no means original; it very much follows the general structure of a standard proof that $\equiv$ and $\cong_\omega$ correspond in classical logic. This latter result is known as the *Ehrenfeucht–Fraïssé criterion*, cf. Fraïssé (1955), Ehrenfeucht (1957) and Feferman (1960).

To simplify the proofs, we consider only prenex formulas, i.e., formulas of the form $Q_1x_1 \ldots Q_nx_n\varphi$, where $n \geq 0$ and φ is quantifier free, and each Q_ix_i is either $\exists x_i$ or $\neg\exists x_i\neg$. The following lemma is proved by minimal modification of a standard argument, and is left to the reader. We let $fv(\varphi)$ be the set of variables occurring free in φ.

Lemma.

(i) *If x does not occur in φ, then $\models_3 \exists x\varphi(x/y) \rightleftharpoons \exists y\varphi$.*

(ii) *For every formula φ, there is a prenex formula ψ such that $fv(\varphi) = fv(\psi)$, and $\models_3 \varphi \rightleftharpoons \psi$.*

Lemma. *If $M \widetilde{\ll}_\omega N$, then $M \overline{\ll} N$.*

Since sentences have no free variables, and all I_n in an ω-partial monomorphism $I = \langle I_n \rangle_{0 \leq n < \omega}$ are non-empty, this lemma follows from the next. But first we need a definition of 'quantifier rank':

Definition. *We define the* quantifier rank *of a formula φ, $qrk(\varphi)$, by induction as follows:*

1. $qrk(\beta) = 0$ *if β is atomic.*
2. $qrk(\neg\varphi) = qrk(\varphi)$.
3. $qrk(\varphi \vee \psi) = max(qrk(\varphi), qrk(\psi))$.
4. $qrk(\exists x\varphi) = qrk(\varphi) + 1$.

Lemma. *Suppose* $I : M \widetilde{\ll}_\omega N$. *For all formulas* φ, *all finite sets* $\{x_1, \ldots, x_r\}$ *of variables and all* $q \in I_n$ *where* $n \geq qrk(\varphi)$, *if* $fv(\varphi) \subseteq \{x_1, \ldots, x_r\}$ *then for any* $a_1, \ldots, a_k \in dom(q)$

$$M \models \varphi[a_1, \ldots, a_r]^\odot \quad \textit{implies} \quad N \models \varphi[q(a_1), \ldots, q(a_r)]^\odot.$$

Proof: We prove this by induction on prenex formulas.

- For quantifier free formulas, the lemma follows from the fact that q is a monomorphism from its domain to its range.
- The induction step for negation is trivial.
- To prove the induction step for $\exists$, suppose the lemma holds for φ, and suppose $fv(\exists x\varphi) \subseteq \{x_1, \ldots, x_k\}$. Moreover, let $q \in I_n$ where $n \geq qrk(\exists x\varphi)$. Note that $n - 1 \geq 0$. We do not assume that $fv(\exists x\varphi) = \{x_1, \ldots, x_k\}$, hence x *could* be among $x_1, \ldots, x_k$. But we can assume that this is not so, since we have an infinite supply of new variables, and $\models_3 \exists x\varphi(x/y) \rightleftharpoons \exists y\varphi$ when x does not occur in φ.
  - We first prove the positive part; suppose
    $$M \models (\exists x\varphi)[a_1, \ldots, a_k]^+.$$
    Then
    $$M \models \varphi[a_1, \ldots, a_k, a]^+$$
    for some a. By the conditions on I there exists a $p \in I_{n-1}$ such that $a \in dom(p)$ and $q \subseteq p$. By the induction hypothesis
    $$N \models \varphi[p(a_1), \ldots, p(a_k), p(a)]^+,$$
    so
    $$N \models (\exists x\varphi)[p(a_1), \ldots, p(a_k)]^+,$$
    i.e.,
    $$N \models (\exists x\varphi)[q(a_1), \ldots, q(a_k)]^+.$$
  - For the negative part we make the contrapositive argument; suppose
    $$N \not\models (\exists x\varphi)[q(a_1), \ldots, q(a_k)]^-.$$
    Then
    $$N \not\models \varphi[q(a_1), \ldots, q(a_k), b]^-$$
    for some b. By the conditions on I there exists a $p \in I_{n-1}$ such that $b \in rng(p)$ and $q \subseteq p$. By the induction hypothesis
    $$M \not\models \varphi[a_1, \ldots, a_k, p^{-1}(b)]^-,$$
    so
    $$M \not\models (\exists x\varphi)[a_1, \ldots, a_k]^-. \qquad \square$$

In order to prove the converse result of this lemma, we first need an auxiliary lemma which describes a limit on the expressibility of formulas built up with certain limited resources. There are just finitely many non-strongly equivalent formulas of a certain quantifier rank, with free variables from a certain finite set. For any $n \geq 0$ and finite set X of variables, let Θ_n^X be the set of formulas φ for which $fv(\varphi) \subseteq X$ and $qrk(\varphi) \leq n$. Each Θ_n^X will be infinite, but the next lemma states that all but a finite number of its elements are indistinguishable semantically:

Lemma. *For any $n \geq 0$ and finite X there is a finite subset Θ_0 of Θ_n^X such that for every $\varphi \in \Theta_n^X$ there is a $\psi \in \Theta_0$ such that $\models_3 \varphi \rightleftharpoons \psi$.*

Proof: $(\Theta_n^X)^*$ and $(\Theta_n^X)_*$ are defined as $\{\varphi^* \mid \varphi \in \Theta_n^X\}$ and $\{\varphi_* \mid \varphi \in \Theta_n^X\}$. The maps * and $_*$ preserve quantifier rank, and $fv(\varphi) = fv(\varphi^*) = fv(\varphi_*)$. Hence by a well known result of classical logic there are just finitely many equivalence classes of classically equivalent formulas in $(\Theta_n^X)^*$, and similarly in $(\Theta_n^X)_*$. By the behavior of * and $_*$, there are just finitely many equivalence classes of positively equivalent formulas in Θ_n^X, and finitely many equivalence classes of negatively equivalent formulas in Θ_n^X. Hence the number of equivalence classes of strongly equivalent (i.e., both positively equivalent and negatively equivalent) formulas must be finite as well. □

For the next proof it will be useful to introduce some new notation. We assume some canonical enumeration of the formulas. If $\varphi_1, \ldots, \varphi_n$ are listed according to their order and $\Pi = \{\varphi_1, \ldots, \varphi_n\}$, then $\bigvee\!\!\!\bigvee \Pi$ is the formula $(\varphi_1 \vee \ldots \vee \varphi_n)$. ($\bigvee\!\!\!\bigvee \emptyset$ is $\bot$.) Similarly $\bigwedge\!\!\!\bigwedge \Pi$ is $\neg \bigvee\!\!\!\bigvee \{\neg\pi \mid \pi \in \Pi\}$. '$\bigvee\!\!\!\bigvee_{a \in \Omega} \varphi_a$' and '$\bigwedge\!\!\!\bigwedge_{a \in \Omega} \varphi_a$' should be read as '$\bigvee\!\!\!\bigvee \{\varphi_a \mid a \in \Omega\}$' and '$\bigwedge\!\!\!\bigwedge \{\varphi_a \mid a \in \Omega\}$'.

Lemma. *If $M \lll N$, then $M \mathrel{\widetilde{\lll}}_\omega N$.*

Proof: Let $x_1, \ldots, x_r, \ldots$ be an infinite list of distinct variables. Moreover, for each $n, k \geq 0$ and $\langle a_1, \ldots, a_k \rangle \in |M|^k$ let $I_n^{\langle a_1, \ldots, a_k \rangle}$ be the set of functions q from $\{a_1, \ldots, a_k\}$ into $|N|$ such that $M \models \varphi[a_1, \ldots, a_k]$ implies $N \models \varphi[q(a_1), \ldots, q(a_k)]$ for all φ with $qrk(\varphi) \leq n$ and $fv(\varphi) \in \{x_1, \ldots, x_k\}$. Then let I_n be the union of all $I_n^{\langle a_1, \ldots, a_k \rangle}$ where $k \geq 0$ and $\langle a_1, \ldots, a_k \rangle \in |M|^k$. Finally, let I be $\langle I_n \rangle_{0 \leq n < \omega}$.

Now $\emptyset \in I_n$ for all n, since $M \lll N$. So every I_n is non-empty. We also note that each $q \in \bigcup_n I_n$ is a monomorphism from its domain to its range, for if $q \in I_n^{\langle a_1, \ldots, a_k \rangle}$ then $M \models \varphi[a_1, \ldots, a_k]$ implies $N \models \varphi[q(a_1), \ldots, q(a_k)]$ for all quantifier free formulas φ.

Thus it only remains to show that I has the back and forth properties. Let $q \in I_{n+1}$ and $b \in |N|$. Then $q \in I_{n+1}^{\langle a_1, \ldots, a_k \rangle}$ for some $k \geq 0$ and k-tuple $\langle a_1, \ldots, a_k \rangle$. We want to show that there exists an $a \in |M|$ such that

$q \cup \{\langle a, b\rangle\} \in I_n^{\langle a_1,\ldots,a_k,a\rangle}$. Suppose not. Then for every $a \in |M|$ there is a formula φ_a such that $qrk(\varphi_a) \leq n$ and $fv(\varphi_a) \in \{x_1, \ldots, x_{k+1}\}$, and such that

$$M \models \varphi_a[a_1, \ldots, a_k, a]$$

but also

$$N \not\models \varphi_a[q(a_1), \ldots, q(a_k), b].$$

Thus

$$M \models \forall x_{k+1} \bigvee_{a\in|M|} \varphi_a[a_1, \ldots, a_k]$$

and

$$N \not\models \bigvee_{a\in|M|} \varphi_a[q(a_1), \ldots, q(a_k), b],$$

i.e.,

$$N \not\models \forall x_{k+1} \bigvee_{a\in|M|} \varphi_a[q(a_1), \ldots, q(a_k)].$$

Since the φ_a all have their free variables among $x_1, \ldots, x_{k+1}$ and all have $qrk \leq n$, we can assume that $\{\varphi_a \mid a \in |M|\}$ is finite. Hence $\forall x_{k+1} \bigvee_{a\in|M|} \varphi_a$ is a proper formula. *However*, since $q \in I_{n+1}^{\langle a_1,\ldots,a_k\rangle}$ and $qrk(\forall x_{k+1} \bigvee_{a\in|M|} \varphi_a) \leq n+1$, we should then, by the definition of $I_{n+1}^{\langle a_1,\ldots,a_k\rangle}$, have

$$N \models \forall x_{k+1} \bigvee_{a\in|M|} \varphi_a[q(a_1), \ldots, q(a_k)],$$

which we have *not*.

To prove the other direction, suppose $q \in I_{n+1}$ and $a \in |M|$. Then $q \in I_{n+1}^{\langle a_1,\ldots,a_k\rangle}$ for some $k \geq 0$ and k-tuple $\langle a_1, \ldots, a_k\rangle$. We want to show that there exists a $b \in |N|$ such that $q \cup \{\langle a, b\rangle\} \in I_n^{\langle a_1,\ldots,a_k,a\rangle}$. Suppose not. Then for every $b \in |N|$ there is a formula φ_b such that $qrk(\varphi_b) \leq n$ and $fv(\varphi_b) \in \{x_1, \ldots, x_{k+1}\}$, and such that

$$M \models \varphi_b[a_1, \ldots, a_k, a]$$

but also

$$N \not\models \varphi_b[q(a_1), \ldots, q(a_k), b].$$

Thus

$$M \models \bigwedge_{b\in|N|} \varphi_b[a_1, \ldots, a_k, a],$$

i.e.,

$$M \models \exists x_{k+1} \bigwedge_{b\in|N|} \varphi_b[a_1, \ldots, a_k];$$

and

$$N \not\models \exists x_{k+1} \bigwedge_{b\in|N|} \varphi_b[q(a_1), \ldots, q(a_k)].$$

Again we can assume that $\exists x_{k+1} \bigwedge_{b \in |N|} \varphi_b$ is a proper formula. This time we can deduce

$$N \models \exists x_{k+1} \bigwedge_{b \in |N|} \varphi_b[q(a_1), \ldots, q(a_k)],$$

which again yields a contradiction. □

We sum up the two directions in one theorem:

Theorem. *$M \widetilde{\lll} N$ iff $M \widetilde{\lll}_\omega N$.*

Hence M and N are semantically equivalent iff there is an ω-partial monomorphism in each direction. From the definitions it is then also immediately seen that $M \equiv N$ if there is an ω-partial isomorphism between them. The reverse does not hold, however; the existence of ω-partial monomorphisms in each direction does not guarantee the existence of one ω-partial *iso*morphism.

Example. *Consider a similarity type with a binary relation symbol R and a unary relation symbol P, and two structures M and N with $|M| = |N| = Z$, the set of integers:*

$$R^{M^+} = R^{N^+} = \{\langle n, n+1 \rangle \mid n \in Z\}$$
$$R^{M^-} = R^{N^-} = \emptyset$$
$$P^{M^+} = \{0, 1, 2, 3 \ldots\}$$
$$P^{N^+} = \{0, 2, 3 \ldots\}$$
$$P^{M^-} = P^{N^-} = \emptyset$$

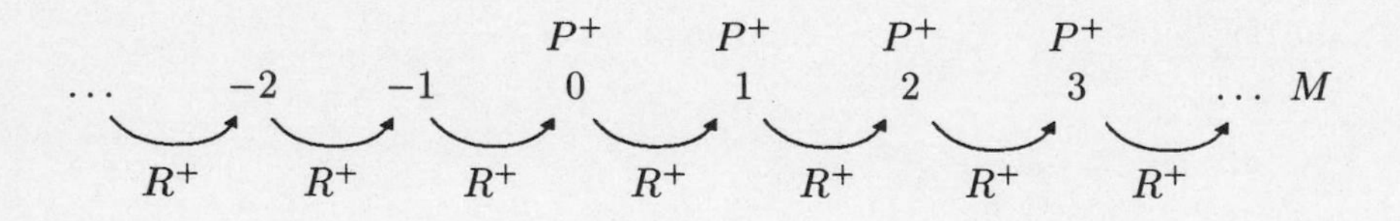

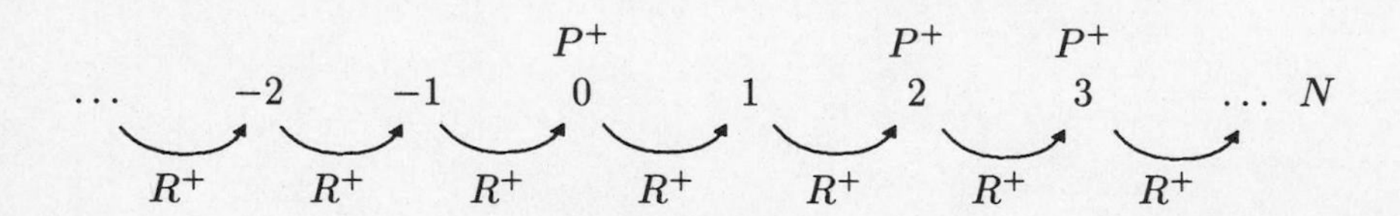

Now $N \lll M$, so clearly $N \widetilde{\lll} M$. On the other hand $g : M \widetilde{\lll} N$, where $g(n) = n + 2$. So there are full monomorphisms in each direction, and

from full monomorphisms ω-partial monomorphisms can always be defined. But there is no ω-partial isomorphism, *since in that case there would be a substructure of M isomorphic to $N \upharpoonright \{0,1,2\}$, which is impossible.*

This failure to distinguish between certain structures that are not ω-partially isomorphic, is connected with persistence. To formulate such a connection we need a general notion of a truth definition. More conditions could be added to the following, but this will be sufficient:

Definition. *A* truth definition *for $\mathcal{L}$ is a relation $\models_x$ between models M and signed formula-assignment pairs $\varphi[A]^{\odot}$, where φ is an $\mathcal{L}$ formula of the similarity type of M. Moreover, $\models_x$ satisfies the following:*

(i) *If $g : M \cong N$ then $M \models_x \varphi[A]^{\odot}$ iff $N \models_x \varphi[g \circ A]^{\odot}$.*

(ii) *If $A(x) = B(x)$ for all x occurring free in φ, then $M \models_x \varphi[A]^{\odot}$ iff $M \models_x \varphi[B]^{\odot}$.*

(iii) *If M is a ρ' model, $\rho \subseteq \rho'$ and $\varphi \in \mathcal{L}[\rho]$, then $M \models_x \varphi[A]^{\odot}$ iff $M \upharpoonright \rho \models_x \varphi[A]^{\odot}$.*

We call these the isomorphism condition, the binding condition and the independence condition. If φ is a sentence and $\models_x$ is a truth definition, we sometimes write $M \models_x \varphi^{\odot}$ if $M \models_x \varphi[A]^{\odot}$ for some A.

Definition. *A truth definition $\models_x$ is* persistent *if for all $\mathcal{L}$ formulas φ, models M and N such that $M \ll N$, and variable assignment A into $|M|$, $M \models_x \varphi[A]^{\odot}$ implies $N \models_x \varphi[A]^{\odot}$.*

Definition. *Let $\models_x$ be a truth definition for $\mathcal{L}$. M and N are* equivalent relative to $\models_x$, *$M \equiv_x N$, if $M \models_x \varphi^{\odot} \leftrightarrow N \models_x \varphi^{\odot}$ for all sentences φ. If it is not the case that $M \equiv_x N$, then $\models_x$* distinguishes *between M and N.*

We can now express more generally the phenomenon observed in the example above:

Theorem. *Let $\models_x$ be a truth definition for $\mathcal{L}$. If $\models_x$ distinguishes between every pair of structures that are not ω-partially isomorphic, then $\models_x$ is not persistent.*

Proof: We use the example above. Let ϕ range over signed sentences. If $\models_x$ distinguishes between M and N, then it is not the case that

$$\forall\phi(M \models_x \phi \leftrightarrow N \models_x \phi),$$

so a ϕ exists such that either

$$M \models_x \phi \quad \text{and} \quad N \not\models_x \phi$$

or

$$M \not\models_x \phi \quad \text{and} \quad N \models_x \phi.$$

If we suppose, for a reductio ad absurdum, that $\models_x$ is persistent, then we can eliminate the latter alternative since $N \ll M$. But since $M \widetilde{\ll} N$, there exists a structure O such that $M \ll O$ and $O \cong N$. Since $M \models_x \phi$, by persistence it follows that $O \models_x \phi$. Since $N \not\models_x \phi$, this contradicts the isomorphism condition on $\models_x$. □

2.6 The Skolem Property

A standard proof of a Skolem–Löwenheim theorem for first order logic uses the notion of *Skolem functions.* In a later chapter we shall take special interest in a version of the Skolem–Löwenheim theorem where counterparts to the Skolem functions are mentioned explicitly. We shall say that a truth definition satisfies the *Skolem property* if this version of the theorem is true for the truth definition. Our counterparts to the Skolem functions are the countable neighborhood assignments:

Definition. *For a given model M, a* countable neighborhood assignment, *abbreviated* cna, *is a function ξ from the finite subsets of $|M|$ to countable subsets of $|M|$, satisfying the following:*

1. $D \subseteq \xi(D)$
2. *If* $D_0 \subseteq D_1$ *then* $\xi(D_0) \subseteq \xi(D_1)$

We say that a subset E of $|M|$ is *closed* under ξ if $\xi(D) \subseteq E$ for all finite subsets D of E:

Definition. *A truth definition $\models_x$ has the* Skolem property *if for every model M there exists a countable neighborhood assignment ξ such that for any non-empty subset E of $|M|$ closed under ξ, any formula φ and variable assignment A into E: $M \models_x \varphi[A]^\odot$ iff $M \upharpoonright E \models_x \varphi[A]^\odot$.*

We observe that the Skolem property implies a familiar version of the Skolem–Löwenheim theorem:

Theorem. *Suppose $\models_x$ has the Skolem property. Then for every structure M there is a countable substructure M_0 such that for all formulas φ and variable assignments A into $|M_0|$, $M \models_x \varphi[A]^\odot$ iff $M_0 \models_x \varphi[A]^\odot$.*

Proof: This follows immediately, since for any model M and *cna* ξ, the closure under ξ of any countable subset D of $|M|$is countable. □

Theorem. $\models$ *has the Skolem property.*

Proof: This follows by a standard argument from classical logic, using Skolem functions. We briefly review the proof, since some of the concepts will be useful later. The argument is formulated slightly different from the standard version, to make it more comparable to a later proof that also $\models\!\!=$ has the Skolem property.

Let the structure M be given, and let g^+ and g^- be functions from pairs $\exists x\varphi[A]$ to elements of M, satisfying

(i) if $M \models \exists x\varphi[A]^+$ then $M \models \varphi[A(g^+(\exists x\varphi[A])/x)]^+$.

(ii) if $M \models \varphi[A(g^-(\exists x\varphi[A])/x)]^-$ then $M \models \exists x\varphi[A]^-$.

Such functions exist by the definition of $\models$ and the axiom of choice. We say that a variable assignment A is *rooted* in an element a, $a \lhd A$, if $A(x) = a$ for all but finitely many variables x. Now let a be an arbitrary element of $|M|$. From g^+, g^- and a we define a function ξ from finite subsets of $|M|$ to subsets of $|M|$. We shall prove that ξ is a *cna*. For any finite set $D \subseteq |M|$, let $\xi(D)$ be the union of the following sets.

D

$\{a\}$

$\{g^+(\exists x\varphi[A]) \mid a \lhd A, rng(A) \subseteq D \cup \{a\}\}$

$\{g^-(\exists x\varphi[A]) \mid a \lhd A, rng(A) \subseteq D \cup \{a\}\}$

Here, 'A' ranges over variable assignments and '$\exists x\varphi$' ranges over all formulas beginning with an existential quantifier. Now ξ is a countable neighborhood assignment. To see this, first note that for any finite set D there are just countably many variable assignments into $D \cup \{a\}$ that are rooted in a. Moreover, there are just countably many formulas. Hence there are just countably many pairs $\exists x\varphi[A]$ where $rng(A) \subseteq D \cup \{a\}$ and $a \lhd A$. Hence we see from the definition of ξ that $\xi(D)$ is countable for any finite D. It should be obvious that ξ also satisfies the two additional conditions on countable neighborhood assignments.

To see that ξ has the property described in the theorem, let E be a nonempty subset of $|M|$ that is closed under ξ. Then $g^{\odot}(\exists x\varphi[A]) \in E$ whenever $a \lhd A$ and $rng(A) \subseteq E$. We prove by induction on the construction of formulas φ that for any A into E, rooted in a,

(*) $M \models \varphi[A]^{\odot}$ iff $M \uparrow E \models \varphi[A]^{\odot}$

From this, the full claim of the theorem will follow from the fact that $\models$ satisfies the binding property. The basis follows directly from the meaning of $\uparrow$ and the definition of $\models$. The validity of the sentential induction steps is read directly off the corresponding clauses in the definition of $\models$. To prove the induction step for $\exists$ we review four separate arguments, corresponding to two directions each of the claims for truth and falsity. Prior to all the four arguments, we assume that $rng(A) \subseteq E$, $a \lhd A$ and that φ satisfies $(*)$.

- Suppose $M \models \exists x\varphi[A]^+$. Let b be $g^+(\exists x\varphi[A])$. Then $M \models \varphi[A(b/x)]^+$, $a \lhd A(b/x)$ and $rng(A(b/x)) \subseteq E$. Hence by the induction hypothesis $M \uparrow E \models \varphi[A(b/x)]^+$ and therefore $M \uparrow E \models \exists x\varphi[A]^+$.

- Suppose $M \uparrow E \models \exists x\varphi[A]^+$. Then $M \models \varphi[A(b/x)]^+$ for some $b \in E$. $a \lhd A(b/x)$, so by the induction hypothesis $M \models \varphi[A(b/x)]^+$ and therefore $M \models \exists x\varphi[A]^+$.

- Suppose $M \uparrow E \models \exists x\varphi[A]^-$. Let b be $g^-(\exists x\varphi[A])$. Now $b \in E$, hence $M \uparrow E \models \varphi[A(b/x)]^-$. Since $a \lhd A(b/x)$ and $rng(A(b/x)) \subseteq E$, by the induction hypothesis $M \models \varphi[A(b/x)]^-$, and therefore $M \models \exists x\varphi[A]^-$ by the condition on g^-.

- Suppose $M \models \exists x\varphi[A]^-$. Then in particular $M \models \varphi[A(b/x)]^-$ for all $b \in E$. Since $a \lhd A(b/x)$, by the induction hypothesis $M \uparrow E \models \varphi[A(b/x)]^-$ for all $b \in E$, hence $M \uparrow E \models \exists x\varphi[A]^-$. $\square$

3

Alternative Truth Definitions

In the first chapter we considered three ways to strengthen the strong Kleene truth definition, and concluded that in propositional logic they all give rise to one and the same truth definition. In this chapter we extend the study to corresponding truth definitions for predicate logic, and we shall discover that the three no longer coincide; in fact they are all distinct: $\models^{wscl}$ is one reasonable counterpart to $\models^{scl}$ in predicate logic; we prove that $\models^{wscl} \leq \models\!\!=^{*}$ and $\models\!\!=^{*} \leq \models_{\Box}$, but that equality holds in neither case.

3.1 The Generalized Strong Kleene Truth Definition

We start with a study of the generalized strong Kleene truth definition. Let 'Γ' range over *finite* sets of signed pairs $\varphi[A]^{\odot}$ of $\mathcal{L}[\rho_M]$ formulas and variable assignments. Similarly, 'R' ranges over elements of ρ_M. $\models\!\!=$ is the smallest relation satisfying the following:

$M \models\!\!= \Gamma, R(x_1,\ldots,x_n)[A]^+, R(y_1,\ldots,y_n)[B]^-$ if $A(x_i) = B(y_i)$ for all i.

$M \models\!\!= \Gamma, R(y_1,\ldots,y_n)[A]^+$ if $\langle A(y_1),\ldots,A(y_n)\rangle \in R^{M^+}$.

$M \models\!\!= \Gamma, R(y_1,\ldots,y_n)[A]^-$ if $\langle A(y_1),\ldots,A(y_n)\rangle \in R^{M^-}$.

$M \models\!\!= \Gamma, (x = y)[A]^+$ if $A(x) = A(y)$.

$M \models\!\!= \Gamma, (x = y)[A]^-$ if $A(x) \neq A(y)$.

$M \models\!\!= \Gamma, \neg\varphi[A]^+$ if $M \models\!\!= \Gamma, \varphi[A]^-$.

$M \models\!\!= \Gamma, \neg\varphi[A]^-$ if $M \models\!\!= \Gamma, \varphi[A]^+$.

$M \models\!\!= \Gamma, (\varphi \vee \psi)[A]^+$ if $M \models\!\!= \Gamma, \varphi[A]^+, \psi[A]^+$.

$M \models\!\!= \Gamma, (\varphi \vee \psi)[A]^-$ if $M \models\!\!= \Gamma, \varphi[A]^-$ and $M \models\!\!= \Gamma, \psi[A]^-$.

$M \models\!\!= \Gamma, \exists x\varphi[A]^+$ if $M \models\!\!= \Gamma, \varphi[A(a/x)]^+$ for some $a \in |M|$.

$M \models\!\!= \Gamma, \exists x\varphi[A]^-$ if $M \models\!\!= \Gamma, \varphi[A(a/x)]^-$ for all $a \in |M|$.

As before, $\models\!\!=^*$ is the restriction of $\models\!\!=$ to single signed pairs of formulas and variable assignments. Hence $M \models\!\!=^* \varphi[A]^\odot$ iff $M \models\!\!= \varphi[A]^\odot$. The *generalized strong Kleene truth definition* is this restricted version.

One important difference between $\models$ and $\models\!\!=$ has to do with "reversibility." We know that if $M \models \exists x\varphi[A]^+$, then $M \models \varphi[A(a/x)]^+$ for some a. On the other hand, $M \models\!\!= \exists x\varphi[A]^+$ might be a consequence of $M \models\!\!= \exists x\varphi[A]^+, \varphi[A(a/x)]^+$ rather than $M \models\!\!= \varphi[A(a/x)]^+$. Thus while the truth of a formula, in the sense of the strong Kleene truth definition, can be "verified" by a procedure that will terminate after a number of steps at most equal to the operator depth of the formula, the same is not true for this generalization of the strong Kleene truth definition. For one and the same Γ, the "effort" it takes to arrive at $M \models\!\!= \Gamma$ can depend critically on the model M. In order to measure this "effort" we shall introduce a notion of *verification trees*, to be made precise below. We use the following set-theoretic definition of a *tree*:

Definition. *For any set D, the set of* trees on D *is the smallest set satisfying the following:*

1. *Every element of D is a tree on D.*
2. *If $a \in D$ and Ω is a set of trees on D, then the ordered pair $\langle a, \Omega \rangle$ is a tree on D.*

By a *tree of "Ds"* we shall mean a tree on D. We assume that the notions of *root, leaf, child node, subtree,* etc., are understood.

With this definition, a given element of D can appear at more than one position ("node") in a tree.

A *verification tree* relative to M is a tree of finite sets Γ of signed pairs $\varphi[A]^\odot$, where the leaves satisfy one of the first five conditions in the definition of $\models\!\!=$, and the remaining nodes are related to their child nodes in a manner that matches one of the remaining conditions in the definition of $\models\!\!=$. If Γ is the set in the root node of a verification tree F, we say that F is a verification tree *for* Γ. Clearly $M \models\!\!= \Gamma$ iff there is a verification tree for Γ in M.

Definition. *We define the* depth $dh(F)$ *of a verification tree F as follows:*

1. *If F is a set Γ then $dh(F) = 0$.*
2. *If $F = \langle a, \Omega \rangle$, where Ω is a set of trees, then $dh(F) = sup\{dh(F_0) \mid F_0 \in \Omega\} + 1$.*

We sometimes write $M \models\!\!=_\alpha \Gamma$ to indicate the existence of a verification tree for Γ of depth at most α in M. Hence we shall use 'α', 'β', etc., to range over *either* ordinals or atomic formulas, as indicated by context.

Lemma. *$M \models_\alpha \Gamma$ for some ordinal α iff $M \models \Gamma$.*

Proof: Since $M \models \Gamma$ iff there is a verification tree for Γ in M, the direction towards the right is trivial. If Γ has a verification tree of depth α then Γ *has* a verification tree.

Similarly, the implication towards the left is equivalent to the assertion that every verification tree has a depth. This is guaranteed by the definition of dh, and the fact that the supremum of any *set* of ordinals is itself an ordinal. □

Lemma. *If $M \models_\alpha \Gamma$, then $M \models_0 \Gamma$ or $M \models_\beta \Gamma$ for some successor ordinal $\beta \leq \alpha$.*

Proof: This is obvious, since the depth of any verification tree is either 0 or a successor ordinal. □

It is natural to ask here whether infinite ordinal depths can yield anything that finite ones cannot; will it follow from $M \models \Gamma$ that $M \models_n \Gamma$ for some finite n? We shall prove that this is not always the case; the full relation $\models$ is stronger than its restriction to finite depths. Before we reach this result, however, we need some simple results about $\models_\alpha$.

Theorem.

(i) *If $M \cong N$, then $M \models_\alpha \Gamma$ iff $N \models_\alpha \Gamma$.*

(ii) *If M is a ρ' model, $\rho \subseteq \rho'$ and only $\mathcal{L}[\rho]$ formulas occur in Γ, then $M \models_\alpha \Gamma$ iff $M \upharpoonright \rho \models_\alpha \Gamma$.*

(iii) *If $A(x) = B(x)$ for all x occurring free in φ, then $M \models_\alpha \Gamma, \varphi[A]^\odot$ iff $M \models_\alpha \Gamma, \varphi[B]^\odot$.*

(iv) *If $A(x) = A(y)$, and z has no free occurrences in φ inside the scope of a quantifier binding x or y, then $M \models_\alpha \Gamma, \varphi(x/z)[A]^\odot$ iff $M \models_\alpha \Gamma, \varphi(y/z)[A]^\odot$.*

The proofs are left to the reader. The two latter we call the *binding property* and the *substitution property*, respectively. Let $\models^*_\alpha$ be the analogue of $\models^*$, i.e., the restriction of $\models_\alpha$ to single signed pairs $\varphi[A]^\odot$. Hence

$$M \models^*_\alpha \varphi[A]^\odot \quad \text{iff} \quad M \models_\alpha \varphi[A]^\odot.$$

We recall the definition of *truth definitions* from the previous chapter. As a special case of (i)–(iii) above it follows that $\models^*_\alpha$ is a truth definition for any α. Since $M \models^* \varphi[A]^\odot$ iff $M \models^*_\alpha \varphi[A]^\odot$ for some α, it follows that also $\models^*$ is a truth definition. We observe that $\models_\alpha$ has the *thinning property*:

Lemma. *If* $M \models_\alpha \Gamma$*, then* $M \models_\alpha \Gamma, \Delta$.

This follows by a simple induction on α, and is left to the reader. We now obtain:

Lemma. *If* $M \models_\alpha \Gamma, \varphi[A]^+$*, then* $M \models_{\alpha+1} \Gamma, (\varphi \vee \psi)[A]^+$ *and* $M \models_{\alpha+1} \Gamma, (\psi \vee \varphi)[A]^+$.

Proof: If $M \models_\alpha \Gamma, \varphi[A]^+$, then by the previous lemma we get $M \models_\alpha \Gamma, \varphi[A]^+, \psi[A]^+$. Hence $M \models_{\alpha+1} \Gamma, (\varphi \vee \psi)[A]^+$ and $M \models_{\alpha+1} \Gamma, (\psi \vee \varphi)[A]^+$. The second of these follows as easily as the first, since 'Γ, $\varphi[A]^+$, $\psi[A]^+$' denotes a *set*. □

In view of the above result, observe that if

- $M \models \Gamma, (\varphi \vee \psi)[A]^+$ if $M \models \Gamma, \varphi[A]^+$
- $M \models \Gamma, (\varphi \vee \psi)[A]^+$ if $M \models \Gamma, \psi[A]^+$

were added as primitive rules, then $\models$ would be the weakening of $\models$ where only a single signed pair $\varphi[A]^\odot$ is allowed at any node in a verification tree. The next lemma reflects this. First we need a definition; for all φ let c_φ be the *operator depth* of φ:

$c_\alpha = 0$ for atomic formulas α.
$c_{\neg\varphi} = c_\varphi + 1$.
$c_{(\varphi \vee \psi)} = max(c_\varphi, c_\psi) + 1$.
$c_{\exists x \varphi} = c_\varphi + 1$.

Lemma. *If* $M \models \varphi[A]^\odot$*, then* $M \models_{c_\varphi} \varphi[A]^\odot$.

The proof is straightforward.

Among the various $\models_\alpha$, the relation $\models_\omega$ holds particular interest. Since no verification tree has depth exactly ω, $M \models_\omega \Gamma$ holds iff Γ has a verification tree in M of finite depth. The efforts in the rest of this section and in the next are directed towards two characterization results, one for each of $\models^*$ and $\models^*_\omega$. We show that the truth definitions $\models^*$ and $\models^*_\omega$ represent the closures of the strong Kleene truth definition under first order inferences in two different senses.

To prove that $\models^*$ is closed under classical first order consequence, we show that

$$\varphi \models_2 \psi \quad \text{implies} \quad M \models \varphi[A]^-, \psi[A]^+,$$

and we show a version of *modus ponens*:

$$\text{If} \quad M \models \varphi[A]^-, \psi[A]^+ \quad \text{and} \quad M \models \varphi[A]^+, \quad \text{then} \quad M \models \psi[A]^+.$$

More generally stated, the latter result takes the form of a derived *cut rule*, quite analogous to the cut rule of a Gentzen style sequent calculus. We prove that

$$\text{if } M \models_\alpha \Gamma, \varphi[A]^+ \text{ and } M \models_\beta \Delta, \varphi[A]^-, \text{ then } M \models \Gamma, \Delta.$$

The proof will follow a standard pattern. The next lemma expresses a special case. We have already used '$\odot$' as a variable ranging over $+$ and $-$. Let $\otimes$ be the opposite of $\odot$, hence '$\otimes$' and '$\odot$' are not independent variables.

Lemma. *If* $M \models_\alpha \Gamma, \varphi[A]^\odot$ *and* $M \models_0 \Delta, \varphi[A]^\otimes$, *then* $M \models \Gamma, \Delta$.

Proof: By induction on α. We can assume that α is a successor ordinal. In each case we assume that $M \models_\alpha \Gamma, \varphi[A]^\odot$ and $M \models_0 \Delta, \varphi[A]^\otimes$, and prove $M \models \Gamma, \Delta$ from the induction hypothesis.

- If $\alpha = 0$, then clearly $M \models \Gamma, \Delta$. This is left to the reader.
- If $\alpha > 0$ and φ is not atomic, then $M \models_0 \Delta$. Hence $M \models_0 \Gamma, \Delta$.
- If $\alpha > 0$ and φ is atomic, then the last step in some verification tree is of the form

$$\frac{\{M \models_{\alpha-1} \Gamma_\vartheta, \varphi[A]^\odot\}_{\vartheta\in\Omega}}{M \models_\alpha \Gamma, \varphi[A]^\odot.}$$

  By the induction hypothesis, $M \models \Gamma_\vartheta, \Delta$ for all $\vartheta \in \Omega$; hence $M \models \Gamma, \Delta$. □

The full cut rule is proved by induction on the sum of α and β. In the argument we need strict monotonicity of "α plus β" in *both* α and β. Hence we cannot use ordinary ordinal addition; instead we shall use the *natural sum* $\alpha \natural \beta$, cf. Levy (1979), p. 130. Observe the following properties of $\natural$.

(i) $\alpha \natural \beta = \beta \natural \alpha$

(ii) $\beta < \beta' \leftrightarrow \alpha \natural \beta < \alpha \natural \beta'$

We also need the ordering $\prec$ on pairs $\langle n, \alpha\rangle$ of natural numbers and ordinals, according to "leftmost difference":

$$\langle n, \alpha\rangle \prec \langle n', \alpha'\rangle \quad \text{iff} \quad \begin{array}{ll} \text{(i)} & n < n', \text{ or} \\ \text{(ii)} & n = n' \text{ and } \alpha < \alpha'. \end{array}$$

Theorem. *If* $M \models_\alpha \Gamma, \varphi[A]^\odot$ *and* $M \models_\beta \Delta, \varphi[A]^\otimes$, *then* $M \models \Gamma, \Delta$.

Proof: By induction on $\langle c_\varphi, \alpha \natural \beta \rangle$, ordered by $\prec$. Suppose $M \models_\alpha \Gamma, \varphi[A]^\odot$ and $M \models_\beta \Delta, \varphi[A]^\otimes$, and suppose the theorem holds for all M', α', β', Γ', Δ', φ', A' such that $\langle c_{\varphi'}, \alpha' \natural \beta' \rangle \prec \langle c_\varphi, \alpha \natural \beta \rangle$. We can assume that α and β are both successor ordinals, hence $\alpha - 1$ and $\beta - 1$ are defined.

- If $\alpha = 0$ or $\beta = 0$, then $M \models \Gamma, \Delta$ by the previous lemma.

- Next suppose $\alpha > 0$, $\beta > 0$ and for some verification tree for $M \models_\alpha \Gamma, \varphi[A]^\odot$, the signed pair $\varphi[A]^\odot$ was not involved in the last step. Then the last step was of the form

$$\frac{\{M \models_{\alpha-1} \Gamma_\vartheta, \varphi[A]^\odot\}_{\vartheta \in \Omega}}{M \models_\alpha \Gamma, \varphi[A]^\odot.}$$

  By the induction hypothesis,

$$M \models \Gamma_\vartheta, \Delta$$

  for all $\vartheta \in \Omega$. Hence

$$M \models \Gamma, \Delta.$$

- The final possibility now is that $\alpha > 0$, $\beta > 0$ and that $\varphi[A]^\odot$ and $\varphi[A]^\otimes$ were both involved in the last steps of their respective verification trees. By symmetry we may assume that $\odot = +$ and $\otimes = -$.

  - First suppose φ is of the form $\neg\varphi_0$. The last step of the verification tree for $M \models_\alpha \Gamma, \neg\varphi_0[A]^+$ corresponds to one of

(i)
$$\frac{M \models_{\alpha-1} \Gamma, \varphi_0[A]^-}{M \models_\alpha \Gamma, \neg\varphi_0[A]^+}$$

(ii)
$$\frac{M \models_{\alpha-1} \Gamma, \neg\varphi_0[A]^+, \varphi_0[A]^-}{M \models_\alpha \Gamma, \neg\varphi_0[A]^+}$$

    By thinning, if the last step was of the form (i), then there is also a verification tree for $M \models \Gamma, \neg\varphi_0[A]^+$ of depth at most α, where the last step was of the form (ii). Hence we may assume that the last step was of this form. By the analogous argument we may assume that the last step of the other verification tree corresponds to

$$\frac{M \models_{\beta-1} \Delta, \neg\varphi_0[A]^-, \varphi_0[A]^+}{M \models_\beta \Delta, \neg\varphi_0[A]^-}$$

Applying the induction hypothesis twice, that is, to the pairs $\langle c_\varphi, (\alpha - 1) \natural \beta \rangle$ and $\langle c_\varphi, \alpha \natural (\beta - 1) \rangle$, we obtain

$$M \models \Gamma, \Delta, \varphi_0[A]^- \text{ and}$$

$$M \models \Gamma, \Delta, \varphi_0[A]^+.$$

Since $c_{\varphi_0} < c_\varphi$, we can apply the induction hypothesis a final time, and obtain

$$M \models \Gamma, \Delta.$$

- Next suppose φ is of the form $(\varphi_0 \vee \varphi_1)$. As in the previous case we may assume that the last steps correspond to

$$\frac{M \models_{\alpha-1} \Gamma, (\varphi_0 \vee \varphi_1)[A]^+, \varphi_0[A]^+, \varphi_1[A]^+}{M \models_\alpha \Gamma, (\varphi_0 \vee \varphi_1)[A]^+}$$

and

$$\frac{\{M \models_{\beta-1} \Gamma, (\varphi_0 \vee \varphi_1)[A]^-, \varphi_i[A]^-\}_{i \in \{0,1\}}}{M \models_\beta \Gamma, (\varphi_0 \vee \varphi_1)[A]^-.}$$

Applying the induction hypothesis once to $\langle c_\varphi, (\alpha - 1) \natural \beta \rangle$ and twice to $\langle c_\varphi, \alpha \natural (\beta - 1) \rangle$, we obtain

$$M \models \Gamma, \Delta, \varphi_0[A]^+, \varphi_1[A]^+$$

$$M \models \Gamma, \Delta, \varphi_0[A]^-$$

$$M \models \Gamma, \Delta, \varphi_1[A]^-.$$

By yet another application of the induction hypothesis we get

$$M \models \Gamma, \Delta, \varphi_0[A]^+,$$

and hence finally

$$M \models \Gamma, \Delta.$$

- Finally suppose φ is of the form $\exists x \varphi_0$. As in the previous cases we may assume that the last steps correspond to

$$\frac{M \models_{\alpha-1} \Gamma, \exists x \varphi_0[A]^+, \varphi_0[A(b/x)]^+}{M \models_\alpha \Gamma, \exists x \varphi_0[A]^+}$$

and

$$\frac{\{M \models_{\beta-1} \Delta, \exists x \varphi_0[A]^-, \varphi_0[A(a/x)]^-\}_{a \in |M|}}{M \models_\beta \Delta, \exists x \varphi_0[A]^-.}$$

By two applications of the induction hypothesis, we obtain

$$M \models \Gamma, \Delta, \varphi_0[A(b/x)]^+ \text{ and}$$

$$M \models \Gamma, \Delta, \varphi_0[A(b/x)]^-.$$

Hence by a final application

$$M \models \Gamma, \Delta. \qquad \square$$

A similar theorem applies for $\models \omega$, and can be proved by a method similar to the above. Such a proof is considerably more complicated, however, since we need to keep track of depths throughout the argument. Hence we shall use a different strategy for $\models_\omega$.

With a cut rule at our disposal, we can now prove that $\models^*_\alpha$ is coherent. It is also persistent, and determinable for infinite α. We sum up:

Theorem.

(i) *If* $M \models \varphi[A]^+$, *then* $M \not\models \varphi[A]^-$.

(ii) *If* M *is complete then either* $M \models_{c_\varphi} \varphi[A]^+$ *or* $M \models_{c_\varphi} \varphi[A]^-$.

(iii) *If* $M \ll N$ *and* $M \models_\alpha \Gamma$, *then* $N \models_\alpha \Gamma$.

Proof: (i) Suppose otherwise. Then by the cut rule $M \models \emptyset$. But this is impossible, since from the first five clauses it is seen that $M \not\models_0 \emptyset$, and all the remaining clauses take verification trees with non-empty roots to verification trees with non-empty roots.

(ii) This is obvious, since $M \models_{c_\varphi} \varphi[A]^\odot$ if $M \models \varphi[A]^\odot$.

(iii) Suppose $M \ll N$. We prove by induction on F that if F is a verification tree relative to M, then F is also a verification tree relative to N.

It is immediately seen that if Γ satisfies one of the first five clauses relative to M, then it satisfies the same clause relative to N. Hence the claim follows for verification trees of depth 0.

Suppose F is a verification tree relative to M, of depth > 0. Then F is of the form $\langle \Gamma, \Omega \rangle$, where Ω is a set of verification trees relative to M. By the induction hypothesis Ω is also a set of verification trees relative to N. We already know that Γ and the roots from Ω satisfy some appropriate mother–child relation. Hence $\langle \Gamma, \Omega \rangle$ is also a verification tree relative to N.

$\square$

Hence $\models^*$ and $\models^*_\omega$ are coherent, determinable and persistent. We now obtain:

Lemma. *The strong and generalized strong Kleene truth definitions coincide on complete structures.*

Proof: We already know that $M \models \varphi[A]^{\odot}$ implies $M \mathrel{|\!\!=\!\!=} \varphi[A]^{\odot}$ for any M. Now let M be complete, and suppose $M \mathrel{|\!\!=\!\!=} \varphi[A]^{\odot}$. By coherence $M \mathrel{|\!\!\neq\!\!=} \varphi[A]^{\otimes}$, and hence $M \not\models \varphi[A]^{\otimes}$. By determinability of $\models$ we get $M \models \varphi[A]^{\odot}$. $\Box$

As before, we define:

$$M \models_{\Box} \varphi[A]^{\odot} \quad \text{iff} \quad (N \models \varphi[A]^{\odot} \quad \text{for all complete } N \text{ such that } M \ll N).$$

Theorem. $\mathrel{|\!\!=\!\!=}^{*} \leq \models_{\Box}$.

Proof: Suppose $M \mathrel{|\!\!=\!\!=} \varphi[A]^{\odot}$, and let N be a completion of M. By persistence $N \mathrel{|\!\!=\!\!=} \varphi[A]^{\odot}$. Hence $N \models \varphi[A]^{\odot}$ by the previous lemma. Since N was arbitrary, $M \models_{\Box} \varphi[A]^{\odot}$. $\Box$

To prove the next theorem we use the following Gentzen-type formal system, which axiomatizes the consequence relation of classical first order logic. It is taken from Barwise (1977), but the notation is somewhat altered. We use 'Φ' and 'Ψ' to range over finite sets of signed formulas:

AXIOM1 $\quad \langle \Phi, \varphi^{\odot}, \varphi^{\otimes} \rangle$

AXIOM2 $\quad \langle \Phi, (x = x)^{+} \rangle$

EQ1 $\quad \dfrac{\langle \Phi, \varphi(x/z)^{\odot} \rangle}{\langle \Phi, \varphi(y/z)^{\odot}, (x = y)^{-} \rangle}$

EQ2 $\quad \dfrac{\langle \Phi, \varphi(x/z)^{\odot} \rangle}{\langle \Phi, \varphi(y/z)^{\odot}, (y = x)^{-} \rangle}$

NEG $\quad \dfrac{\langle \Phi, \varphi^{\odot} \rangle}{\langle \Phi, \neg\varphi^{\otimes} \rangle}$

DIS^{+} $\quad \dfrac{\langle \Phi, \varphi^{+}, \psi^{+} \rangle}{\langle \Phi, (\varphi \vee \psi)^{+} \rangle}$

DIS^{-} $\quad \dfrac{\langle \Phi, \varphi^{-} \rangle \qquad \langle \Phi, \psi^{-} \rangle}{\langle \Phi, (\varphi \vee \psi)^{-} \rangle}$

EXIST $\quad \dfrac{\langle \Phi, \varphi(y/x)^{\odot} \rangle}{\langle \Phi, \exists x \varphi^{\odot} \rangle}$

In the EQ rules, z must not occur free in φ inside the scope of any quantifier binding x or y. In EXIST, x must not occur free in φ inside the scope of a quantifier binding y. Moreover, if $\odot = -$, then y must not occur free in the resulting sequent.

In AXIOM1, φ can be any formula. However, it is easy to see that the set of derivable sequents is the same if in AXIOM1 we only allow φ to be atomic. In the following we shall assume this constraint.

Now $\langle\Phi\rangle$ iff $\models_2 \Phi$, i.e., iff for each complete M and A, $M \models \varphi[A]^{\odot}$ for some $\varphi^{\odot} \in \Phi$.

Analogously to the notation '$M \models\!\!=_n \Gamma$', we shall also write '$\langle\Phi\rangle_n$' to indicate the existence of a derivation of $\langle\Phi\rangle$ of depth at most n. When Φ is a set of signed formulas, $\Phi[A] = \{\varphi[A]^{\odot} \mid \varphi^{\odot} \in \Phi\}$.

Lemma. *If* $\langle\Phi, \Psi\rangle_n$ *and* $M \models \varphi[A]^{\otimes}$ *for all* $\varphi^{\odot} \in \Phi$, *then* $M \models\!\!=_n \Psi[A]$.

A different formulation of the lemma lends itself to a shorter proof. For any Φ, M and A, let $\{\Phi\}_{M,A}$ be the set $\{\varphi^{\odot} \in \Phi \mid M \not\models \varphi[A]^{\otimes}\}$. By thinning, the next lemma is equivalent to the above:

Lemma. *If* $\langle\Phi\rangle_n$, *then* $M \models\!\!=_n \{\Phi\}_{M,A}[A]$.

Proof: By induction on n.

- If $\langle\Phi\rangle_0$, then $M \models\!\!=_0 \Phi_{M,A}[A]$. This is left to the reader.
- Let the following be an instance of EQ1.

$$\frac{\langle\Phi, \varphi(x/z)^{\odot}\rangle_{n-1}}{\langle\Phi, \varphi(y/z)^{\odot}, (x = y)^{-}\rangle_n.}$$

  Moreover, suppose

$$M \models\!\!=_{n-1} \{\Phi, \varphi(x/z)^{\odot}\}_{M,A}[A].$$

  - If $A(x) = A(y)$, then $M \models\!\!=_{n-1} \{\Phi, \varphi(y/z)^{\odot}\}_{M,A}[A]$, since both $\models$ and $\models\!\!=_{n-1}$ satisfy the substitution property. Hence also $M \models\!\!=_n \{\Phi, \varphi(y/z)^{\odot}, (x = y)^{-}\}_{M,A}[A]$.
  - If $A(x) \neq A(y)$, then $M \models (x = y)[A]^{-}$, $M \not\models (x = y)[A]^{+}$, and $(x = y)^{-} \in \{\Phi, \varphi(y/z)^{\odot}, (x = y)^{-}\}_{M,A}$. Hence trivially $M \models\!\!=_n \{\Phi, \varphi(y/z)^{\odot}, (x = y)^{-}\}_{M,A}[A]$.

- The step for EQ2 is similar. The steps for disjunction and negation are left to the reader.
- Let the following be an instance of EXIST.

$$\frac{\langle\Phi, \varphi(y/x)^{+}\rangle_{n-1}}{\langle\Phi, \exists x\varphi^{+}\rangle_n.}$$

  Moreover, suppose

$$M \models\!\!=_{n-1} \{\Phi, \varphi(y/x)^{+}\}_{M,A}[A].$$

- If $M \models \varphi(y/x)[A]^-$, then $\{\Phi, \varphi(y/x)^+\}_{M,A} \subseteq \{\Phi, \exists x\varphi^+\}_{M,A}$, and we are done.
- If $M \not\models \varphi(y/x)[A]^-$, then $\{\Phi, \varphi(y/x)^+\}_{M,A} = \{\Phi\}_{M,A}, \varphi(y/x)^+$. Since therefore

$$M \models_{n-1} \{\Phi\}_{M,A}[A], \varphi(y/x)[A]^+,$$

and since x is not free in $\varphi(y/x)$, by the binding and substitution properties

$$M \models_{n-1} \{\Phi\}_{M,A}[A], \varphi[A(A(y)/x)]^+.$$

Hence we need only add a step of existential quantification to obtain

$$M \models_{n} \{\Phi\}_{M,A}[A], \exists x\varphi[A]^+.$$

Clearly $M \not\models \exists x\varphi[A]^-$, hence $\{\Phi, \exists x\varphi^+\}_{M,A} = \{\Phi\}_{M,A}, \exists x\varphi^+$, and we are done.

- Let the following be an instance of EXIST.

$$\frac{\langle\Phi, \varphi(y/x)^-\rangle_{n-1}}{\langle\Phi, \exists x\varphi^-\rangle_n}$$

where y does not occur free in Φ or $\exists x\varphi$. Moreover, suppose

$$M \models_{n-1} \{\Phi, \varphi(y/x)^-\}_{M,B}[B]$$

for all B, and let A be an arbitrary variable assignment.

- If $M \models \exists x\varphi[A]^+$, then $M \models \exists y(\varphi(y/x))[A]^+$ and hence $M \models \varphi(y/x)[A(a/y)]^+$ for some a. Hence $\{\Phi, \varphi(y/x)^-\}_{M,A(a/y)} = \{\Phi\}_{M,A(a/y)}$, and therefore by assumption $M \models_{n-1} \{\Phi\}_{M,A(a/y)} [A(a/y)]$. Since y does not occur free in Φ, also $M \models_{n-1} \{\Phi\}_{M,A}[A]$. Hence clearly $M \models_n \{\Phi, \exists x\varphi^-\}_{M,A}[A]$.
- If $M \not\models \exists x\varphi[A]^+$, then $\{\Phi, \exists x\varphi^-\}_{M,A} = \{\Phi\}_{M,A}, \exists x\varphi^-$. By assumption and possibly thinning,

$$M \models_{n-1} \{\Phi\}_{M,A(a/y)}[A(a/y)], \varphi(y/x)[A(a/y)]^-.$$

Since y does not occur free in Φ, by the binding property we have

$$M \models_{n-1} \{\Phi\}_{M,A}[A], \varphi(y/x)[A(a/y)]^-.$$

Since x does not occur free in $\varphi(y/x)$, by the binding and substitution properties we have

$$M \models_{n-1} \{\Phi\}_{M,A}[A], \varphi[A(a/y)(a/x)]^-.$$

From this we may conclude

$$M \models_{n-1} \{\Phi\}_{M,A}[A], \varphi[A(a/x)]^-.$$

(If $x = y$, this is trivial. If $x \neq y$, then y does not occur free in φ since it does not occur free in $\exists x\varphi$, and so the conclusion follows by the binding property.) Since this holds for all $a \in |M|$, we need only add a step of negative existential quantification to obtain

$$M \models_n \{\Phi\}_{M,A}[A], \exists x\varphi[A]^-. \qquad \square$$

The next theorem is an immediate consequence. By notational convention introduced above, 'Φ' and 'Ψ' range over *finite* sets of signed formulas. A stronger version for infinite sets will follow by compactness of the classical consequence relation.

Theorem. *If* $\Phi \models_2 \Psi$, *and* $M \models \varphi[A]^{\odot}$ *for all* $\varphi^{\odot} \in \Phi$, *then* $M \models_{\omega} \Psi[A]$.

Corollary. *If* $\Phi \models_2 \varphi^{\odot}$, *and* $M \models^* \varphi_0[A]^{\odot_0}$ *for all* $\varphi_0^{\odot_0} \in \Phi$, *then* $M \models^* \varphi[A]^{\odot}$.

Proof: If $\Phi \models_2 \varphi^{\odot}$, i.e., $\models_2 \{\varphi_0^{\otimes_0} \mid \varphi_0^{\odot_0} \in \Phi\} \cup \{\varphi^{\odot}\}$, then $M \models \{\varphi_0[A]^{\otimes_0} \mid \varphi_0^{\odot_0} \in \Phi\} \cup \{\varphi[A]^{\odot}\}$. If also $M \models^* \varphi_0[A]^{\odot_0}$ for all $\varphi_0^{\odot_0} \in \Phi$, then $M \models^* \varphi[A]^{\odot}$ by repeated applications of the derived cut rule. $\square$

Hence $\models^*$ is closed under classical consequence. In the next section we prove a similar result for $\models^*_{\omega}$.

3.2 Syntactic Closures

There are at least two reasonable generalizations of $\models^{scl}$ to predicate logic. It turns out that one coincides with $\models^*$, the other with $\models^*_{\omega}$.

Let $\models^{sscl}$, the *strong syntactic closure* of $\models$, be defined as the smallest relation satisfying the following.

1. If $M \models \varphi[A]^{\odot}$ then $M \models^{sscl} \varphi[A]^{\odot}$.
2. If $\Phi \models_2 \varphi^{\odot}$ and $M \models^{sscl} \varphi_0[A]^{\odot_0}$ for all $\varphi_0^{\odot_0} \in \Phi$, then $M \models^{sscl} \varphi[A]^{\odot}$.
3. If $M \models^{sscl} \varphi[A]^+$ and $A(x) = B(x)$ for all x free in φ, then $M \models^{sscl} \varphi[B]^+$.
4. If $M \models^{sscl} \varphi[A(a/x)]^-$ for all $a \in |M|$, then $M \models^{sscl} \exists x\varphi[A]^-$.

Here, '$\varphi^{\odot}$' ranges over signed $\mathcal{L}[\rho_M]$ formulas, and 'Φ' ranges over sets of such signed formulas. Since the consequence relation $\models_2$ is compact, it is immaterial whether we restrict to *finite* sets.

Similarly we let $\models^{wscl}$, the *weak* syntactic closure of $\models$ be the smallest relation satisfying 1 and 2, but not necessarily 3 or 4. (It is, however, an easy exercise to check that $\models^{wscl}$ *will* satisfy 3 as well.)

Note that a positive counterpart to 4 would be superfluous in the definition of $\models^{sscl}$: If $M \models^{sscl} \varphi[A(a/x)]^+$ for some a, then $M \models^{sscl} \exists x\varphi[A(a/x)]$ since $\varphi \models_2 \exists x\varphi$. And hence $M \models^{sscl} \exists x\varphi[A]^+$ by condition 3.

We shall prove that $\models^{sscl} = \models\!\!=^*$ and $\models^{wscl} = \models\!\!=^*_\omega$. Hence it will follow that $\models^{sscl}$ and $\models^{wscl}$ are *truth definitions*, i.e., they satisfy the isomorphism, binding and independence conditions. The first two are obvious, but the third is a fairly deep result. It has the flavor of an interpolation result, but it is not obvious how to derive it from any existing result of this type. Until the independence property is established we shall be careful in our arguments, and keep in mind that for all that is proved so far $M \models^{sscl} \varphi[A]^{\odot}$ could depend crucially on the similarity type of M. But note that $\models^{sscl}$ is still *well defined*, since with our notion of models the similarity type of M is uniquely determined.

Theorem. *$\models^{sscl} \leq \models\!\!=^*$ and $\models^{wscl} \leq \models\!\!=^*_\omega$.*

Proof: We know that $\models\!\!=^*$ satisfies conditions 1–4 in the definition of $\models^{sscl}$, hence $\models^{sscl}$ is contained in $\models\!\!=^*$.

To prove $\models^{wscl} \leq \models\!\!=^*_\omega$, suppose $M \models^{wscl} \varphi[A]^{\odot}$. A straightforward argument by induction on the definition of $\models^{wscl}$ shows that in this case $M \models \chi[A]^+$ and $\chi \models_2 \varphi^{\odot}$ for some χ. But then $M \models\!\!=_\omega \varphi[A]^{\odot}$ follows by the last theorem of the previous section. □

To prove the opposite result for $\models^{wscl}$ we show that if $M \models\!\!=_\omega \varphi[A]$, then $M \models \chi[A]$ for a χ classically equivalent to φ.

First consider an example. If $P^{M^+} = P^{M^-} = \emptyset$, then

$$M \not\models \exists x \forall y(P(x) \vee \neg P(y))[A]^+,$$

while

$$M \models\!\!=_{12} \exists x \forall y(P(x) \vee \neg P(y))[A]^+.$$

Here, as always, $\forall y$ is short for $\neg\exists y \neg$. The verification tree has leaf nodes of the form

$$M \models\!\!=_0 P(x)[A_{a,b}]^+, P(y)[A_{a,b}]^-, P(x)[A_{b,c}]^+, P(y)[A_{b,c}]^-,$$

where $A_{a,b} = A(a/x)(b/y)$ and $A_{b,c} = A(b/x)(c/y)$. Hence $A_{a,b}(y) = A_{b,c}(x)$. After several steps, we arrive at

$$M \models\!\!=_8 (P(x) \vee \neg P(y))[A_{a,b}]^+, \exists x \forall y(P(x) \vee \neg P(y))[A]^+$$

for all a and b. Four more steps yield

$$M \models_{12} \exists x \forall y (P(x) \lor \neg P(y))[A]^+, \exists x \forall y (P(x) \lor \neg P(y))[A]^+,$$

i.e.,

$$M \models_{12} \exists x \forall y (P(x) \lor \neg P(y))[A]^+.$$

The "i.e." corresponds to the *contraction rule* of a sequent calculus that uses sequences rather than sets. Let φ be $\exists x \forall y (P(x) \lor \neg P(y))$. The possibility of really satisfying a "multiple" rather than a single occurrence of $\varphi[A]^+$, is part of what makes $\models$ and $\models_\omega$ stronger than $\models$. Without resort to contraction we would not get $M \models \varphi[A]^+$, while we *would* get $M \models_{13} (\varphi \lor \varphi)[A]^+$. Also a prenex form of this multiple is satisfied without resort to contraction:

$$M \models_{13} \exists x \forall y \exists x' \forall y' ((P(x) \lor \neg P(y)) \lor (P(x') \lor \neg P(y')))[A]^+.$$

A little more processing gives us a sentence φ', classically equivalent to φ, such that $M \models \varphi'[A]^+$.

φ' will have twice the quantifier rank of φ. When more instances of contraction are needed on a single path in the tree, then the quantifier rank may multiply even more. The depth of the verification tree puts a bound on this multiplication of quantifier rank, but with infinite depths we may get a formula of infinite quantifier rank, which is of course not an $\mathcal{L}$ formula at all. That is why the result can be proved for $\models_\omega$ and not for $\models$.

We now turn to the general argument. To carry out the induction we need to state a more general result about finite sets Φ of signed formulas. We generalize the definition of *quantifier rank* correspondingly: $qrk(\Phi) = max\{qrk(\varphi) \mid \varphi^\odot \in \Phi\}$. Similarly, $fv(\Phi) = \cup_{\varphi^\odot \in \Phi} fv(\varphi)$. $\bigvee\!\!\!\bigvee \Phi$ is the formula $\bigvee\!\!\!\bigvee \{\varphi \mid \varphi^+ \in \Phi\} \lor \bigvee\!\!\!\bigvee \{\neg\varphi \mid \varphi^- \in \Phi\}$.

Lemma. *If* $M \models_n \Phi[A]$, *then* $M \models \chi[A]$ *for a formula* χ *such that*

(i) $qrk(\chi) \leq n + qrk(\Phi)$

(ii) $fv(\chi) \subseteq fv(\Phi)$

(iii) $\models_3 \neg\chi \equiv \neg \bigvee\!\!\!\bigvee \Phi$

Proof: By induction on n. We follow the clauses in the definition of $\models$.

- Suppose $R(x_1, \ldots, x_m)^+ \in \Phi$ and $R(y_1, \ldots, y_m)^- \in \Phi$, where $A(x_i) = A(y_i)$ for all i: $1 \leq i \leq m$.
  Then let χ be $\bigvee\!\!\!\bigvee \Phi \lor ((x_1 = y_1) \land \ldots \land (x_m = y_m))$. By the assumption on A and on the x_i and y_i we have

$$M \models ((x_1 = y_1) \land \ldots \land (x_m = y_m))[A],$$

so $M \models \chi[A]$. Clearly χ satisfies (i) and (ii). To see that it satisfies (iii), it is sufficient to show

$$\models_3 \neg(\neg R(x_1, \ldots, x_m) \vee R(y_1, \ldots, y_m)) \supset \neg((x_1 = y_1) \wedge \ldots \wedge (x_m = y_m)).$$

But this is a consequence of the coherence condition on structures.

- If $M \models\!\!=_0 \Phi[A]$ by one of the four remaining basic clauses, let χ be $\bigvee\!\!\!\!\bigvee \Phi$ itself.
- In both the positive and negative step for negation, we can use the same χ given by the induction hypothesis.
- In the positive step for disjunction we can use the same χ given by the induction hypothesis.
- For the negative step for disjunction, suppose $\Phi = \Phi_0, (\varphi_0 \vee \varphi_1)^-$ and the last step of some verification tree is of the form

$$\frac{M \models\!\!=_{n-1} \Phi_0[A], \varphi_0[A]^- \qquad M \models\!\!=_{n-1} \Phi_0[A], \varphi_1[A]^-}{M \models\!\!=_n \Phi_0[A], (\varphi_0 \vee \varphi_1)[A]^-.}$$

By the induction hypothesis there are χ_0 and χ_1 such that $M \models \chi_0[A]$ and $M \models \chi_1[A]$, where χ_0 satisfies (i)–(iii) relative to Φ_0, φ_0^- and χ_1 satisfies (i)–(iii) relative to Φ_0, φ_1^-. Now $M \models (\chi_0 \wedge \chi_1)[A]$, and it is easy to check that $(\chi_0 \wedge \chi_1)$ satisfies (i)–(iii) relative to $\Phi_0, (\varphi_0 \vee \varphi_1)^-$.

- For positive existential quantification, suppose $\Phi = \Phi_0, \exists x\varphi^+$ and the last step of some verification tree is of the form

$$\frac{M \models\!\!=_{n-1} \Phi_0[A], \varphi[A(a/x)]^+}{M \models\!\!=_n \Phi_0[A], \exists x\varphi[A]^+.}$$

Let y be a variable not occurring in either of Φ_0 or $\exists x\varphi$. Using the binding and substitution properties several times, we deduce

$$M \models\!\!=_{n-1} \Phi_0[A(a/y)], \varphi(y/x)[A(a/y)]^+.$$

Hence by the induction hypothesis $M \models \chi_0[A(a/y)]$ for a χ_0 such that

(1) $qrk(\chi_0) \leq n - 1 + qrk(\Phi_0, \varphi(y/x)^+)$

(2) $fv(\chi_0) \subseteq fv(\Phi_0, \varphi(y/x)^+)$

(3) $\models_3 \neg\chi_0 \equiv \neg(\bigvee\!\!\!\!\bigvee \Phi_0 \vee (\varphi(y/x)))$

Hence also $M \models \exists y\chi_0[A]$. For $\exists y\chi_0$ we deduce

(1′) $qrk(\exists y\chi_0) \le n + qrk(\Phi_0, \exists x\varphi^+)$

(2′) $fv(\exists y\chi_0) \subseteq fv(\Phi_0, \exists x\varphi^+)$

(3′) $\models_3 \neg\exists y\chi_0 \equiv \neg\exists y(\bigvee\!\!\!\bigvee \Phi_0 \vee (\varphi(y/x)))$

But since y did not occur in either of φ or Φ_0, it then follows that also

(3″) $\models_3 \neg\exists y\chi_0 \equiv \neg(\bigvee\!\!\!\bigvee \Phi_0 \vee \exists x\varphi)$.

- For negative existential quantification, suppose $\Phi = \Phi_0, \exists x\varphi^-$ and the last step of some verification tree is of the form

$$\frac{\{M \models_{n-1} \Phi_0[A], \varphi[A(a/x)]^-\}_{a\in|M|}}{M \models_n \Phi_0[A], \exists x\varphi[A]^-.}$$

Let y be a variable not occurring in either of Φ_0 or $\exists x\varphi$. Using the binding and substitution properties several times, we deduce

$$M \models_{n-1} \Phi_0[A(a/y)], \varphi(y/x)[A(a/y)]^-$$

for every $a \in |M|$. Hence by the induction hypothesis $M \models \chi_a[A(a/y)]$ for a χ_a such that

(1) $qrk(\chi_a) \le n - 1 + qrk(\Phi_0, \varphi(y/x)^-)$

(2) $fv(\chi_a) \subseteq fv(\Phi_0, \varphi(y/x)^-)$

(3) $\models_3 \neg\chi_a \equiv \neg(\bigvee\!\!\!\bigvee \Phi_0 \vee (\neg\varphi(y/x)))$

Since each χ_a satisfies (1) and (2), all but finitely many of them are strongly equivalent, and we may assume that all but finitely many of them are identical. Hence $\bigvee\!\!\!\bigvee_{a\in|M|}\chi_a$ is a formula, let χ_0 be this formula. We now have $M \models \chi_0[A(a/y)]$ for each $a \in |M|$, and hence $M \models \forall y\chi_0[A]$. For $\forall y\chi_0$ we deduce

(1′) $qrk(\forall y\chi_0) \le n + qrk(\Phi_0, \exists x\varphi^-)$

(2′) $fv(\forall y\chi_0) \subseteq fv(\Phi_0, \exists x\varphi^-)$

(3′) $\models_3 \neg\forall y\chi_0 \equiv \neg\forall y(\bigvee\!\!\!\bigvee \Phi_0 \vee (\neg\varphi(y/x)))$

But since y did not occur in φ or Φ_0, it then follows that also

(3″) $\models_3 \neg\forall y\chi_0 \equiv \neg(\bigvee\!\!\!\bigvee \Phi_0 \vee \neg\exists x\varphi)$. □

We have proved:

Theorem. *If* $M \models_\omega \varphi[A]$ *then* $M \models \chi[A]$ *for a* χ *such that* $\models_3 \neg\varphi \equiv \neg\chi$.

Corollary. $\models^{wscl} = \models^*_\omega$.

Proof: By the behavior of the negation sign, it is sufficient to prove that $M \models_\omega \varphi[A]^+$ iff $M \models^{wscl} \varphi[A]^+$. If $\models_3 \neg\varphi \equiv \neg\chi$ then $\models_2 \varphi \equiv \chi$, so if $M \models_\omega \varphi[A]^+$ then $M \models^{wscl} \varphi[A]^+$. The other direction is already proved. □

We proceed towards a proof that $\models^{sscl}$ is identical to $\models$. The following lemma is central:

Lemma. *If* $M \models_\alpha \Phi[A]$, *then* $M \models^{sscl} \bigvee\Phi[A]$.

Proof: By induction on α. We can assume that α is 0 or a successor ordinal; in the latter case $\alpha - 1$ is defined. We follow the clauses in the definition of $\models$.

- Clearly, if $M \models_0 \Phi[A]$, then $M \models^{sscl} \bigvee\Phi[A]$.

- In both the positive and negative step for negation, as well as the positive step for disjunction, the lemma is a trivial application of the induction hypothesis since there is virtually no change in the corresponding formula $\bigvee\Phi$.

- For the negative step for disjunction, suppose $\Phi = \Phi_0, (\varphi_0 \vee \varphi_1)^-$ and the last step of some verification tree is of the form

$$\frac{M \models_{\alpha-1} \Phi_0[A], \varphi_0[A]^- \qquad M \models_{\alpha-1} \Phi_0[A], \varphi_1[A]^-}{M \models_\alpha \Phi_0[A], (\varphi_0 \vee \varphi_1)[A]^-.}$$

  By the induction hypothesis

$$M \models^{sscl} \bigvee\Phi_0 \vee \neg\varphi_0[A] \quad \text{and} \quad M \models^{sscl} \bigvee\Phi_0 \vee \neg\varphi_1[A],$$

  hence

$$M \models^{sscl} \bigvee\Phi_0 \vee \neg(\varphi_0 \vee \varphi_1)[A].$$

- For positive existential quantification, suppose $\Phi = \Phi_0, \exists x\varphi^+$ and the last step of some verification tree is of the form

$$\frac{M \models_{\alpha-1} \Phi_0[A], \varphi[A(a/x)]^+}{M \models_\alpha \Phi_0[A], \exists x\varphi[A]^+}$$

Let y be a variable not occurring in either of Φ_0 or $\exists x\varphi$. Using the binding and substitution properties several times, we can deduce

$$M \models_{\alpha-1} \Phi_0[A(a/y)], \varphi(y/x)[A(a/y)]^+.$$

Hence by the induction hypothesis

$$M \models^{sscl} \bigvee \Phi_0 \vee (\varphi(y/x))[A(a/y)].$$

Since

$$\bigvee \Phi_0 \vee (\varphi(y/x)) \models_2 \bigvee \Phi_0 \vee \exists x\varphi,$$

we now get

$$M \models^{sscl} \bigvee \Phi_0 \vee \exists x\varphi[A(a/y)].$$

Hence by clause 3 in the definition of $\models^{sscl}$,

$$M \models^{sscl} \bigvee \Phi_0 \vee \exists x\varphi[A].$$

- For negative existential quantification, suppose $\Phi = \Phi_0, \exists x\varphi^-$ and the last step of some verification tree is of the form

$$\frac{\{M \models_{\alpha-1} \Phi_0[A], \varphi[A(a/x)]^-\}_{a\in|M|}}{M \models_{\alpha} \Phi_0[A], \exists x\varphi[A]^-.}$$

Let y be a variable not occurring in Φ_0 or φ. For every $a \in |M|$ we have

$$M \models_{\alpha-1} \Phi_0[A(a/y)], \varphi(y/x)[A(a/y)]^-.$$

By the induction hypothesis,

$$M \models^{sscl} \bigvee \Phi_0 \vee \neg(\varphi(y/x))[A(a/y)]^+.$$

Hence

$$M \models^{sscl} \neg\bigvee \Phi_0 \wedge (\varphi(y/x))[A(a/y)]^-$$

for each $a \in |M|$. Hence by clause 4 in the definition of $\models^{sscl}$

$$M \models^{sscl} \exists y(\neg\bigvee \Phi_0 \wedge (\varphi(y/x)))[A]^-.$$

Since y does not occur in Φ_0 or φ, by clause 2 also

$$M \models^{sscl} \bigvee \Phi_0 \vee \neg\exists x\varphi[A]. \qquad \square$$

As a special case of the above, it follows that $M \models \varphi[A]$ implies $M \models^{sscl} \varphi[A]$. Hence by the behavior of the negation sign, $\models^* \leq \models^{sscl}$, and we have proved:

Theorem. $\models^{sscl} = \models^*$.

3.3 Neighborhood Assignments

Up until now, we have shown several *positive* results about the relations between truth definitions. Specifically, we have shown:

$$\models^{wscl} = \models\!\!=^*_\omega \leq \models^{sscl} = \models\!\!=^* \leq \models_\Box.$$

In this and the next section we prove two *negative* results; we give examples to show that $\models\!\!=^*_\omega \neq \models\!\!=^*$ and $\models\!\!=^* \neq \models_\Box$. Both of these results build on results about *neighborhood assignments*:

Definition. *For a given model M, a neighborhood assignment is a function ξ from the finite subsets of $|M|$ to subsets of $|M|$, satisfying the following:*

1. $D \subseteq \xi(D)$.
2. *If* $D_0 \subseteq D_1$ *then* $\xi(D_0) \subseteq \xi(D_1)$.

Given a neighborhood assignment ξ, we define a corresponding restriction $\models_\xi$ of the truth definition $\models$. $\models_\xi$ is defined by an induction exactly matching the definition of $\models$, except in the inductive clauses for $\exists$:

- $M \models_\xi \exists x\varphi[A]^+$ iff $M \models_\xi \varphi[A(a/x)]^+$ for some $a \in \xi(\{A(x_1),\ldots, A(x_n)\})$, where $x_1,\ldots,x_n$ are the free variables of $\exists x\varphi$.
- $M \models_\xi \exists x\varphi[A]^-$ iff $M \models_\xi \varphi[A(a/x)]^-$ for every $a \in |M|$.

So the negative clauses are the same. Notice that $\models_\xi$ may create gaps of its own, not stemming from gaps at the atomic level. It turns out that $\models_\xi$ will in some cases violate the condition of determinability. But it will still be a useful tool. The lack of determinability makes it possible for $\models_\xi$ to have a nice monotonicity property, as stated in the next lemma. We define $\xi \leq \zeta$ to hold iff $\xi(D) \subseteq \zeta(D)$ for every finite D:

Lemma. *If $\xi \leq \zeta$, then $M \models_\xi \varphi[A]^\odot$ implies $M \models_\zeta \varphi[A]^\odot$.*

Proof: By induction on φ.

It is sufficient to consider the positive step for $\exists$. Suppose $M \models_\xi \exists x\varphi[A]^+$. Then $M \models_\xi \varphi[A(a/x)]^+$ for some $a \in \xi(A[fv(\exists x\varphi)])$. By the induction hypothesis $M \models_\zeta \varphi[A(a/x)]^+$. Since $\xi \leq \zeta$, also $a \in \zeta(A[fv(\exists x\varphi)])$, so $M \models_\zeta \exists x\varphi[A]^+$. □

There is a maximal neighborhood assignment ξ_{max} which maps every finite subset to the whole domain itself. Clearly $\models$ and $\models_{\xi_{max}}$ are the same. Hence the next lemma immediately follows from the previous.

Lemma. *If $M \models_\xi \varphi[A]^\odot$, then $M \models \varphi[A]^\odot$.*

We have already defined a countable neighborhood assignment (*cna*) to be a neighborhood assignment ξ for which $\xi(D)$ is countable for each finite D. We now define:

Definition. *$M \models_{\Box\xi} \varphi[A]^{\odot}$ iff $N \models_{\xi} \varphi[A]^{\odot}$ for all completions N of M. $M \models_{\Box\aleph_0} \varphi[A]^{\odot}$ iff $M \models_{\Box\xi} \varphi[A]^{\odot}$ for some* cna *ξ.*

Since ξ_{max} is a *cna* on countable models, $\models_{\Box\aleph_0}$ and $\models_{\Box}$ coincide on such models. As our main result about $\models_{\Box\aleph_0}$ we shall prove that $M \models_{\Box\aleph_0} \varphi[A]^{\odot}$ always implies $M \models \varphi[A]^{\odot}$. Hence $\models^*$ and $\models_{\Box}$ coincide on countable models:

Theorem. $\models_{\Box\aleph_0} \leq \models^*$.

Proof: This proof is a lot like a standard completeness proof for first order logic, using the method of semantic tableaux. Note that the theorem is equivalent to the following assertion. If $M \not\models \varphi[A]^{\odot}$, then for any *cna* ξ there is a completion M' of M such that $M' \not\models_{\xi} \varphi[A]^{\odot}$.

So suppose that $M \not\models \varphi[A]^{\odot}$, and let ξ be an arbitrary *cna*. Since each $\xi(D)$ is countable, we can suppose that each $\xi(D)$ is given an enumeration. We now define a sequence $\langle \Gamma_n \rangle_{n<\omega}$ of finite sets Γ_n where for each $n < \omega$, $\Gamma_n \subseteq \Gamma_{n+1}$ and $M \not\models \Gamma_n$.

(0) $\Gamma_0 = \{\varphi[A]^{\odot}\}$.

(1) Γ_{5r+1} is defined from Γ_{5r} as follows:

Γ_{5r} (like all sets of signed pairs $\psi[B]^{\odot}$) must be of the form

$$\Delta, (\psi_1^0 \vee \psi_1^1)[A_1]^+, \ldots, (\psi_m^0 \vee \psi_m^1)[A_m]^+$$

where $m \geq 0$ and no element of Δ is of the form $(\psi^0 \vee \psi^1)[B]^+$.

Since

$$M \not\models \Gamma_{5r},$$

we have

$$M \not\models \Gamma_{5r}, \psi_1^0[A_1]^+, \psi_1^1[A_1]^+, \ldots, \psi_m^0[A_m]^+, \psi_m^1[A_m]^+.$$

Let Γ_{5r+1} be this new set.

(2) Γ_{5r+2} is defined from Γ_{5r+1} as follows:

Γ_{5r+1} is of the form

$$\Delta, (\psi_1^0 \vee \psi_1^1)[A_1]^-, \ldots, (\psi_m^0 \vee \psi_m^1)[A_m]^-$$

where $m \geq 0$ and no element of Δ is of the form $(\psi^0 \vee \psi^1)[B]^-$.

Since

$$M \not\models \Gamma_{5r+1},$$

for some tuple $\langle i_1, \ldots, i_m \rangle \in \{0,1\}^m$ we have

$$M \not\models \Gamma_{5r+1}, \psi_1^{i_1}[A_1]^-, \ldots, \psi_m^{i_m}[A_m]^-.$$

Let Γ_{5r+2} be this new set.

(3) Γ_{5r+3} is defined from Γ_{5r+2} as follows:

Γ_{5r+2} is of the form

$$\Delta, \exists x_1 \psi_1[A_1]^+, \ldots, \exists x_m \psi_m[A_m]^+$$

where $m \geq 0$ and no element of Δ is of the form $\exists x \psi[B]^+$.

Since

$$M \not\models \Gamma_{5r+2},$$

we know that

$$M \not\models \Gamma_{5r+2}, \psi_1[A_1(a_1/x_1)]^+, \ldots, \psi_m[A(a_m/x_m)]^+$$

for any sequence $\langle a_1, \ldots, a_m \rangle$.

Let Γ_{5r+3} be

$$\Gamma_{5r+2}, \psi_1[A_1(c_1/x_1)]^+, \ldots, \psi_m[A_m(c_m/x_m)]^+$$

where c_i is the least element e of $\xi(A_i[fv(\exists x_i \psi_i)])$ such that the signed pair $\psi_i[A_i(e/x_i)]^+$ is not an element of Γ_{5r+2}. ('Least' refers to the enumeration mentioned at the beginning of the proof.) If no such e exists, a corresponding new element of Γ_{5r+3} does not have to be added.

(4) Γ_{5r+4} is defined from Γ_{5r+3} as follows:

Γ_{5r+3} is of the form

$$\Delta, \exists x_1 \psi_1[A_1]^-, \ldots, \exists x_m \psi_m[A_m]^-$$

where $m \geq 0$ and no element of Δ is of the form $\exists x \psi[B]^-$.

Since

$$M \not\models \Gamma_{5r+3},$$

we know that

$$M \not\models \Gamma_{5r+3}, \psi_1[A_1(a_1/x_1)]^-, \ldots, \psi_m[A_m(a_m/x_m)]^-$$

for some sequence $\langle a_1, \ldots, a_m \rangle$. Let Γ_{5r+4} be this new set.

(5) Γ_{5r+5} is defined from Γ_{5r+4} as follows:

Γ_{5r+4} is of the form

$$\Delta, \neg\psi_1[A_1]^+, \dots, \neg\psi_m[A_m]^+, \neg\chi_1[B_1]^-, \dots, \neg\chi_k[B_k]^-$$

where $m, k \geq 0$ and no element of Δ is of the form $\neg\psi[B]^+$ or $\neg\chi[B]^-$. Since

$$M \not\models \Gamma_{5r+4},$$

we know that

$$M \not\models \Gamma_{5r+4}, \psi_1[A_1]^-, \dots, \psi_m[A_m]^-, \chi_1[B_1]^+, \dots, \chi_k[B_k]^+.$$

Let Γ_{5r+5} be this new set.

Finally let Γ_ω be $\bigcup_{n<\omega} \Gamma_n$. We observe that $M \not\models \Delta$ for each finite $\Delta \subseteq \Gamma_\omega$. Hence

- There is no pair $R(x_1, \dots, x_n)[B]^+$ and $R(y_1, \dots, y_n)[C]^-$ in Γ_ω such that $B(x_i) = C(y_i)$ for all i: $1 \leq i \leq n$.
- If $R(x_1, \dots, x_n)[B]^+ \in \Gamma_\omega$ then $\langle B(x_1), \dots, B(x_n)\rangle \notin R^{M^+}$.
- If $R(x_1, \dots, x_n)[B]^- \in \Gamma_\omega$ then $\langle B(x_1), \dots, B(x_n)\rangle \notin R^{M^-}$.
- If $(x = y)[B]^+ \in \Gamma_\omega$ then $B(x) \neq B(y)$.
- If $(x = y)[B]^- \in \Gamma_\omega$ then $B(x) = B(y)$.

Moreover, by the definition of Γ_ω it is seen that

- If $\neg\psi[B]^+ \in \Gamma_\omega$ then $\psi[B]^- \in \Gamma_\omega$.
- If $\neg\psi[B]^- \in \Gamma_\omega$ then $\psi[B]^+ \in \Gamma_\omega$.
- If $(\psi_0 \vee \psi_1)[B]^+ \in \Gamma_\omega$ then $\psi_0[B]^+ \in \Gamma_\omega$ and $\psi_1[B]^+ \in \Gamma_\omega$.
- If $(\psi_0 \vee \psi_1)[B]^- \in \Gamma_\omega$ then $\psi_0[B]^- \in \Gamma_\omega$ or $\psi_1[B]^- \in \Gamma_\omega$.
- If $\exists x\psi[B]^+ \in \Gamma_\omega$ then $\psi[B(a/x)]^+ \in \Gamma_\omega$ for all $a \in \xi(B[fv(\exists x\psi)])$.
- If $\exists x\psi[B]^- \in \Gamma_\omega$ then $\psi[B(a/x)]^- \in \Gamma_\omega$ for some a.

Now let M' be any completion of M such that

$$\langle B(x_1), \dots, B(x_n)\rangle \in R^{M'^+} \quad \text{if} \quad R(x_1, \dots, x_n)[B]^- \in \Gamma_\omega,$$

and

$$\langle B(x_1), \dots, B(x_n)\rangle \in R^{M'^-} \quad \text{if} \quad R(x_1, \dots, x_n)[B]^+ \in \Gamma_\omega.$$

By the first three observations such an extension exists. By induction on the subformulas ψ of φ it is now seen that $M' \not\models_\xi \psi[B]^\odot$ if $\psi[B]^\odot \in \Gamma_\omega$. Hence $M' \not\models_\xi \varphi[A]^\odot$, so $M \not\models_{\Box\xi} \varphi[A]^\odot$.

Since ξ was an arbitrary *cna*, also $M \not\models_{\Box\aleph_0} \varphi[A]^\odot$. $\Box$

Corollary. *$\models_\Box$ and $\models^*$ coincide on countable structures.*

We have seen that $\models_{\Box\aleph_0} \leq \models^*$. Equality does not hold, however, since $\models_{\Box\aleph_0}$ is not closed under first order consequence. We leave it for the reader to check that if M has uncountable domain and $P^{M^+} = P^{M^-} = \emptyset$, then $M \not\models_{\Box\aleph_0} \exists x\forall y(P(x) \vee \neg P(y))$, despite the fact that this sentence is a theorem of first order logic.

The extent of the difference between $\models_{\Box\aleph_0}$ and $\models^*$ is not known. For instance, it is an open question whether $\models^*$ is the closure of $\models_{\Box\aleph_0}$ under first order consequence, i.e., whether $M \models^* \varphi[A]$ implies the existence of a formula ψ such that $M \models_{\Box\aleph_0} \psi[A]$ and $\psi \models_2 \varphi$.

We are almost ready to prove that $\models^*$ and $\models^*_\omega$ are distinct, but we need the following theorem:

Theorem. *Let ϕ be a signed sentence. If $M \widetilde{\ll}_\omega N$ and $M \models_\omega \phi$, then $N \models_\omega \phi$.*

Proof: We prove this as a corollary to the corresponding result about $\models$. It is sufficient to consider the case where ϕ is a positively signed sentence φ^+. Suppose $M \widetilde{\ll}_\omega N$, and suppose $M \models_\omega \varphi^+$. Then there exists a formula ψ and a variable assignment A such that $M \models \psi[A]$ and $\psi \models_2 \varphi$. Let τ be the existential closure of ψ. We have $M \models \tau$ and $\tau \models_2 \varphi$. Hence $N \models \tau$, and $N \models^{wscl} \varphi$. $\Box$

Theorem. $\models^* \neq \models^*_\omega$.

Proof: Consider a similarity type containing only a binary relation symbol S. Let $|M|$ be Z, the set of integers. $S^{M^+} = \{\langle n, n+1\rangle \mid n \in Z\}$, $S^{M^-} = \emptyset$. $|N|$ is $Z' \cup Z''$, the union of two disjoint copies of the integers. $S^{N^+} = \{\langle n, n+1\rangle \mid n \in Z'\} \cup \{\langle n, n+1\rangle \mid n \in Z''\}$, $S^{N^-} = \emptyset$.

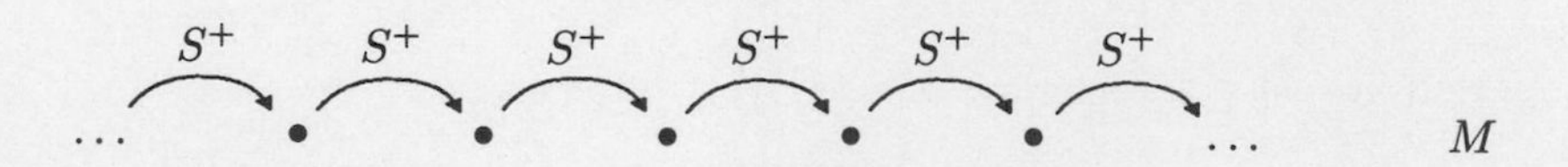

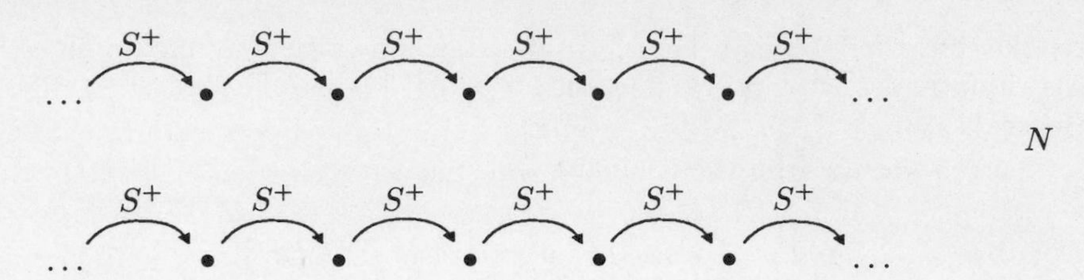

Now let $trans(S)$ and $conn(S)$ be

$$\forall x \forall y \forall z (S(x,y) \land S(y,z) \to S(x,z))$$

and

$$\forall x \forall y (x = y \lor S(x,y) \lor S(y,x))$$

respectively. We have

$$M \models_\Box trans(S) \to conn(S)$$

but not

$$N \models_\Box trans(S) \to conn(S).$$

Since both M and N have countable domains, we also have

$$M \models trans(S) \to conn(S)$$

and

$$N \not\models trans(S) \to conn(S).$$

So $\models^*$ distinguishes between M and N. On the other hand, we know from the above theorem that $\models_\omega$ does not distinguish between ω-partially isomorphic structures. In the following we describe an ω-partial isomorphism from M to N. Hence it will follow that $\models^*_\omega$ does not distinguish between the two structures, and that the two truth definitions $\models^*$ and $\models^*_\omega$ are distinct.

If $e_0 \in Z'$ and $e_1 \in Z''$, let $|e_0 - e_1| = |e_1 - e_0| = \omega$. Otherwise, $-$ and $|\ |$ are ordinary subtraction and absolute value operator. For each $n \geq 0$, let I_n be the set of bijections f from finite subsets of $|M|$ to finite subsets of $|N|$, satisfying

1. If $|a - b| \leq 2^n$ then $a - b = f(a) - f(b)$.

2. If $|a - b| > 2^n$ then $|f(a) - f(b)| > 2^n$.

Clearly each I_n is non-empty, and each element of I_n is an isomorphism from its domain to its range. Let I be $\langle I_n \rangle_{0 \leq n < \omega}$. We prove that $I : M \cong_\omega N$. To see that I has the back and forth properties, suppose $p \in I_{n+1}$ and $a \in |M|$. We show that there is a $b \in |N|$ such that $p \cup \{\langle a, b\rangle\} \in I_n$.

If $|a - c| > 2^n$ for all $c \in dom(p)$, let b be any element of $|N|$ such that $|b - p(c)| > 2^n$ for all $c \in dom(p)$.

Otherwise, if $|a - c| \leq 2^n$ for some $c \in dom(p)$, then let b be the unique element such that $b - p(c) = a - c$. There may be several such c, but by the condition on elements of I_{n+1} it is seen that b is well defined by the above equation. Clearly, $p \cup \{\langle a, b\rangle\} \in I_n$. The other direction is similar. □

3.4 The Skolem Property

In the preceding section we found a pair of structures that were indistinguishable relative to $\models\!=^*_\omega$, but not relative to $\models\!=^*$. Hence $\models\!=^*_\omega \neq \models\!=^*$.

In this section we use a similar strategy for the pair $\models_\Box$, $\models\!=^*$. We again prove that the expressibility properties are different: $\models\!=^*$ satisfies a certain Skolem–Löwenheim theorem, while $\models_\Box$ does not. We recall the definition of the Skolem property:

Theorem. *The generalized strong Kleene truth definition has the Skolem property.*

This is a special case of the next lemma, which is a corresponding result about the full relation $\models\!=$. We define $range(\Gamma)$ as the union of $range(A)$ for all variable assignments A occurring in Γ:

Lemma. *For every model M there is a countable neighborhood assignment ξ such that for any non-empty subset E of M closed under ξ and any Γ such that $range(\Gamma) \subseteq E$, $M \models\!= \Gamma$ iff $M \uparrow E \models\!= \Gamma$.*

Proof: To prove this result we again define Skolem functions or "witness functions" g^+ and g^- as in the proof that $\models$ has the Skolem property. Due to the properties of $\models\!=$, we now have to give these functions two arguments, as described below. But more importantly, we need to be more careful this time about *which* witnesses we pick:

We say that b is a witness *for* Γ and $\exists x\varphi[A]$ in M if $M \models\!= \Gamma, \varphi[A(b/x)]^+$. (Note that even if $M \models\!= \Gamma, \exists x\varphi[A]^+$, there does not necessarily exist any witness for Γ and $\exists x\varphi[A]$ in M. This follows from the lack of "reversibility" noted at the beginning of the chapter.) We say that b is a *minimal witness* for Γ and $\exists x\varphi[A]$ in M if b is a witness, and for any ordinal α and any $c \in |M|$, if $M \models\!=_\alpha \Gamma, \varphi[A(c/x)]^+$ then also $M \models\!=_\alpha \Gamma, \varphi[A(b/x)]^+$. Clearly, if Γ and $\exists x\varphi[A]$ have a witness in M, then they also have a minimal witness.

We recall the definition of a variable assignment A being *rooted* in an element a, $a \lhd A$. We extend this to sets Γ of signed pairs $\varphi[A]^{\odot}$: $a \lhd \Gamma$ iff $a \lhd A$ for each A occurring in Γ.

We again define two witness functions g^+ and g^-. Both are *two-argument* functions defined on pairs Γ and $\exists x\varphi[A]$. If Γ and $\exists x\varphi[A]$ have a witness in M, let $g^+(\Gamma, \exists x\varphi)$ be a *minimal* witness. Similarly, if $M \not\models \Gamma, \exists x\varphi[A]^-$, then there exists an element b of $|M|$ such that $M \not\models \Gamma, \varphi[A(b/x)]^-$. Let $g^-(\Gamma, \exists x\varphi[A])$ be such a b, if one exists. Moreover, let a be some element of $|M|$. From g^+, g^- and a we now define a neighborhood assignment ξ: For any finite set $D \subseteq |M|$, let $\xi(D)$ be the union of the following four sets.

$$D$$

$$\{a\}$$

$$\{g^+(\Gamma, \exists x\varphi[A]) \mid a \lhd \Gamma, a \lhd A, rng(\Gamma) \cup rng(A) \subseteq D \cup \{a\}\}$$

$$\{g^-(\Gamma, \exists x\varphi[A]) \mid a \lhd \Gamma, a \lhd A, rng(\Gamma) \cup rng(A) \subseteq D \cup \{a\}\}$$

Here, '$\exists x\varphi$' ranges over all formulas beginning with an existential quantifier. ξ is clearly a neighborhood assignment. It is also a *countable* neighborhood assignment. To see this, first note that for any finite set D there are just countably many variable assignments B into $D \cup \{a\}$ that are rooted in a. Moreover, there are just countably many formulas ψ. Hence there are just countably many signed pairs $\psi[B]^{\odot}$ where $rng(B) \subseteq D \cup \{a\}$ and $a \lhd B$. Hence also there are just countably many finite sets Γ of such pairs. Hence if D is finite, then there are just countably many pairs $\Gamma, \exists x\varphi[A]$ satisfying the conditions

$$a \lhd \Gamma,\ a \lhd A,\ rng(\Gamma) \cup rng(A) \subseteq D \cup \{a\},\ \exists x\varphi \text{ formula.}$$

Hence we see from the definition of ξ that $\xi(D)$ is countable for any finite D. This proves that ξ is a *cna*. To show that ξ has the property described in the lemma, let E be a non-empty subset of $|M|$ that is closed under ξ.

The remainder of the proof is divided in two parts. We first prove by induction on α that for any Γ such that $rng(\Gamma) \subseteq E$,

$$\text{if} \quad M \models_{\alpha} \Gamma, \quad \text{then} \quad M \uparrow E \models \Gamma.$$

By the binding property, we can assume that all variable assignments are rooted in a. The basis of the induction follows directly from the meaning of '$\uparrow$' and the definition of $\models$. The validity of the sentential induction steps is read directly off the corresponding clauses in the definition of $\models$. In both of the induction steps for $\exists$, we assume that $rng(\Gamma) \subseteq E$ and $rng(A) \subseteq E$.

- Suppose the last step in a verification tree of depth α is of the form

$$\frac{M \models \Gamma, \varphi[A(c/x)]^+}{M \models \Gamma, \exists x\varphi[A]^+.}$$

Let b be $g^+(\Gamma, \exists x\varphi[A])$. Since b is a minimal witness, also $M \models \Gamma, \varphi[A(b/x)]^+$ by a verification tree of depth less than α. $b \in E$, so by the induction hypothesis $M \uparrow E \models \Gamma, \varphi[A(b/x)]^+$ and hence $M \uparrow E \models \Gamma, \exists x\varphi[A]^+$.

- Suppose the last step in a verification tree of depth α is of the form

$$\frac{\{M \models \Gamma, \varphi[A(c/x)]^-\}_{c\in|M|}}{M \models \Gamma, \exists x\varphi[A]^-.}$$

Then by the induction hypothesis $M \uparrow E \models \Gamma, \varphi[A(c/x)]^-$ for all $c \in E$, and therefore $M \uparrow E \models \Gamma, \exists x\varphi[A]^-$.

This proves one direction. For the opposite direction, we prove by induction on α that for any Γ such that $rng(\Gamma) \subseteq E$,

$$\text{if} \quad M \uparrow E \models_\alpha \Gamma, \quad \text{then} \quad M \models \Gamma.$$

By the binding property, we can again assume that all variable assignments are rooted in a. Again, the basis of the induction follows directly from the meaning of '$\uparrow$' and the definition of $\models$. The validity of the sentential induction steps is read directly off the corresponding clauses in the definition of $\models$.

- Suppose the last step in a verification tree of depth α is of the form

$$\frac{M \uparrow E \models \Gamma, \varphi[A(c/x)]^+}{M \uparrow E \models \Gamma, \exists x\varphi[A]^+.}$$

By the induction hypothesis $M \models \Gamma, \varphi[A(c/x)]^+$; hence $M \models \Gamma, \exists x\varphi[A]^+$.

- Suppose the last step in a verification tree of depth α is of the form

$$\frac{\{M \uparrow E \models \Gamma, \varphi[A(c/x)]^-\}_{c\in E}}{M \uparrow E \models \Gamma, \exists x\varphi[A]^-.}$$

Then in particular $M \uparrow E \models \Gamma, \varphi[A(b/x)]^-$ by a verification tree of depth less than α, where $b = g^-(\Gamma, \exists x\varphi[A])$. By the induction hypothesis $M \models \Gamma, \varphi[A(b/x)]^-$; hence by the definition of g^- it follows that $M \models \Gamma, \exists x\varphi[A]^-$. □

We recall that a truth definition $\models_x$ is *at least as strong as* a truth definition $\models_y$, written $\models_y \leq \models_x$, iff $M \models_y \varphi[A]^\odot$ implies $M \models_x \varphi[A]^\odot$ for all M, φ, A and $\odot$. Moreover, $\models_x$ is *reliable* iff $\models_x \leq \models_\Box$. We note the following:

Corollary. *The generalized strong Kleene truth definition is the strongest reliable truth definition with the Skolem property.*

Proof: We already know that $\models^*$ is a reliable truth definition with the Skolem property. We must show that if $\models_x$ is any reliable truth definition with the Skolem property, then $\models_x \leq \models^*$. So let $\models_x$ be such a truth definition, and suppose $M \models_x \varphi[A]^\odot$. Since $\models_x$ and $\models^*$ both have the Skolem property, there exist two countable neighborhood assignments ξ and ζ in M with the appropriate relation to $\models_x$ and $\models^*$ respectively.

The formula φ contains just finitely many variables. By the binding condition on truth definitions we can, without loss of generality, assume that A has finite range. Let M_0 be a countable substructure of M where the domain is a superset of $range(A)$, and closed under both ξ and ζ. Clearly such a countable substructure exists.

Since $M \models_x \varphi[A]^\odot$, also $M_0 \models_x \varphi[A]^\odot$, and hence $M_0 \models_\Box \varphi[A]^\odot$ since $\models_x$ is reliable. Since $\models^*$ and $\models_\Box$ coincide on countable structures, $M_0 \models^* \varphi[A]^\odot$. But then $M \models^* \varphi[A]^\odot$ immediately follows. □

The next objective is to prove that the Skolem property fails for $\models_\Box$. Hence it will follow that $\models_\Box$ and $\models^*$ are distinct. We have defined $M \models_\Box \varphi[A]^\odot$ to hold iff $N \models \varphi[A]^\odot$ for all completions N of M, i.e., for all complete N such that $M \ll N$. There is a sense in which this truth definition is second order: By quantifying over all informational completions of the model, we are in effect quantifying over all the relations completing those referred to by the relation symbols. For instance

$$M \models_\Box \forall x \exists y R(x,y)$$

is really equivalent to

$$M \models \forall R'(complete(R') \wedge extends(R,R') \supset \forall x \exists y R'(x,y))$$

where $\models$ is extended to second-order logic in the obvious way. (The quantification ranges over the possible interpretations for binary relation symbols: Ordered pairs of disjoint subsets of $|M|^2$.) Here, $complete(R')$ and $extends(R,R')$ are the sentences

$$\forall x \forall y (R'(x,y) \vee \neg R'(x,y))$$

and

$$\forall x \forall y(R(x,y) \supset R'(x,y)) \wedge \forall x \forall y(\neg R(x,y) \supset \neg R'(x,y))$$

respectively.

Essentially the same is the case for the supervaluation semantics for free logic. In that context Peter Woodruff shows (cf. Woodruff 1984) how to encode Π^1_1 formulas of classical second order logic in this supervaluation semantics. With such resources certain cardinality facts can be expressed, and thus counterexamples can be found for both the upwards and downwards Skolem–Löwenheim theorems.

Now Woodruff's results can not be transported directly to the truth definition $\models_\Box$ of the present framework. In free logic, "partiality" arises in atomic formulas containing a non-denoting term. In the simplest case the formula $R(c)$ is neither true nor false if c does not denote. This form of partiality or indeterminateness is expressible; the formula $\neg\exists x(x = c)$ says that c does not denote. This feature is essential to Woodruff's argument; we shall see that a simple form of the Skolem–Löwenheim theorem in fact holds for $\models_\Box$.

Theorem. *For any signed sentence ϕ and structure M, if $M \models_\Box \phi$ then $M_0 \models_\Box \phi$ for some countable structure M_0.*

Proof: If $M \models_\Box \phi$, then $M' \models_2 \phi$ for a completion M' of M. Since M' is complete and the language is countable, by the downwards Skolem–Löwenheim theorem for classical first order logic there is a complete model N with countable domain such that $N \models_2 \phi$. Since N is complete it has exactly one completion, namely itself, so $N \models_2 \phi$ and $N \models_\Box \phi$ are equivalent. $\Box$

The above proof is not very illuminating. There are some second-order properties around, but they do not result in a failure of this weak version of the Skolem–Löwenheim theorem because the language contains only persistent operators. For this reason we shall consider also a strengthened version which involves the notion of substructures:

Theorem. *It is not the case that for every signed sentence ϕ and structure M, if $M \models_\Box \phi$, then $M_0 \models_\Box \phi$ for some countable substructure M_0 of M.*

Proof: Consider a similarity type with three relation symbols Q, F and $<$, where Q is unary while F and $<$ are binary. We define the model M:

- $|M|$ is the set of real numbers.
- Q^{M^+} is the rationals.
- Q^{M^-} is the complement of Q^{M^+} in $|M|$, hence Q is totally defined.

- $<^{M^+}$ is the usual strict ordering relation of the reals.
- $<^{M^-}$ is the complement of $<^{M^+}$ in $|M|^2$, hence $<$ is totally defined.
- $F^{M^+} = F^{M^-} = \emptyset$, so F is totally undefined.

Let σ be the conjunction of the two sentences

$$\forall x \exists y (Q(y) \wedge F(y, x))$$

and

$$\forall x \forall y \forall z (F(x, y) \wedge F(x, z) \rightarrow y = z).$$

Here, $\varphi \rightarrow \psi$ is defined as $\neg\varphi \vee \psi$. σ says that F is a function and that F maps Q^{M^+} onto the whole domain of M. This is impossible for cardinality reasons; hence $M' \models_2 \neg\sigma$ for every completion M' of M. Hence $M \models_\Box \neg\sigma$.

Furthermore, let τ be the sentence

$$\forall x \exists y (x < y \wedge Q(y)).$$

$M \models \tau$, hence also $M \models_\Box \tau$ and $M \models_\Box \tau \wedge \neg\sigma$.

We next show that no countable subset $D \subseteq |M|$ exists such that

$$M \uparrow D \models_\Box \tau \wedge \neg\sigma.$$

Suppose the contrary. If $M \uparrow D \models_\Box \tau \wedge \neg\sigma$ then clearly $M \uparrow D \models_\Box \tau$, and thus also $M \uparrow D \models \tau$ since $<$ is totally defined in M and hence in $M \uparrow D$. It follows that D contains infinitely many rationals. Since D is countable there exists therefore a surjective function from the rationals in D to D itself. If N is a completion of $M \uparrow D$ where F is interpreted as such a function, then clearly $N \models \sigma$. Hence $M \uparrow D \not\models_\Box \neg\sigma$, contrary to our assumption. □

Corollary. $\models_\Box \neq \models^*$.

Proof: As a special case of the above theorem, it follows that $\models_\Box$ does not have the Skolem property. □

A little more should be said about $\models_\Box$ and Skolem–Löwenheim properties. For sets Γ and Δ of signed $\mathcal{L}[\rho]$ sentences, we define $\Gamma \models_\Box \Delta$ to hold iff for all ρ models M, if $M \models_\Box \phi$ for all $\phi \in \Gamma$, then $M \models_\Box \phi$ for some $\phi \in \Delta$. In the special case where Δ is a single signed sentence ϕ, it turns out that $\Gamma \models_\Box \phi$ and $\Gamma \models_2 \phi$ are equivalent. The direction towards the right is trivial, since $\models_\Box$ coincides with $\models$ on complete structures. The other direction follows from the fact that $\models_\Box$ is closed under classical consequence. (We shall see later, however, that such an equivalence does not hold for the more general case $\Gamma \models_\Box \phi_0, \phi_1$.)

Now, we have seen that whenever $M \models_\Box \phi$, then trivially $N \models \phi$ for some complete N; hence $N_0 \models \phi$ for some countable, complete N_0; hence $N_0 \models_\Box \phi$. Given the above remarks, we can improve on this. For any M and ϕ_0 such that $M \not\models_\Box \phi_0$, we get $\Gamma_M \not\models_\Box \phi_0$. Here, $\Gamma_M = \{\phi \in \mathcal{L}[\rho] \mid M \models_\Box \phi\}$. Hence $\Gamma_M \not\models_2 \phi_0$, and there is a countable, complete N_0 such that $N_0 \not\models \phi_0$, and $N_0 \models \phi$ for all $\phi \in \Gamma_M$. In other words, for every model M and signed sentence ϕ_0 such that $M \not\models_\Box \phi_0$, there is a countable model M_0 such that $M_0 \not\models_\Box \phi_0$, and for all signed sentences ϕ, if $M \models_\Box \phi$ then $M_0 \models_\Box \phi$.

As noted in the parenthesized remark above, this cannot be strengthened to the slightly stronger version with *two* signed sentences ϕ_0 and ϕ_1, replacing the single signed sentence ϕ_0 above. This deserves mention, since it shows that yet another simple Skolem–Löwenheim property fails for $\models_\Box$, this one not involving the notion of substructures:

Definition. $\models_x$ *has the* Löwenheim property *if for every* ρ *and every* ρ *model* M *there exists a countable model* M_0 *such that for all signed* $\mathcal{L}[\rho]$ *sentences* ϕ, $M \models_x \phi$ *iff* $M_0 \models_x \phi$.

Clearly the Skolem property implies the Löwenheim property. The labeling of these two properties is natural, but should not be taken too seriously. The name of the latter property is chosen merely to distinguish it from the former.

Theorem. $\models_\Box$ *does not have the Löwenheim property.*

Proof: We return to the counterexample used in the the proof of the previous theorem. Let M, Q, F, $<$, σ and τ be as before. In addition, let π be the conjunction of the sentences

$$\forall x \forall y \forall z (x < y \land y < z \to x < z) \quad \text{and} \quad \forall x \neg (x < x).$$

Finally, let φ_0 and φ_1 be

$$\exists x \exists y F(x, y) \quad \text{and} \quad \exists x \exists y \neg F(x, y)$$

respectively. Clearly $M \models_\Box \pi$, $M \not\models_\Box \varphi_0$ and $M \not\models_\Box \varphi_1$. We already know that $M \models_\Box \tau \land \neg\sigma$. Now suppose $\models_\Box$ satisfies the Löwenheim property. Then there is countable model M_0 such that $M_0 \models_\Box \pi \land \tau \land \neg\sigma$, while $M_0 \not\models_\Box \varphi_0$ and $M_0 \not\models_\Box \varphi_1$. Then $F^{M_0^+} = F^{M_0^-} = \emptyset$, and $Q^{M_0^+}$ is infinite. But then clearly there is a completion N of M_0 such that $N \models \sigma$, contradicting the assumption $M_0 \models_\Box \neg\sigma$. □

3.5 Semantical Equivalence for Structures

In this section we identify a sufficient condition that two structures are equivalent relative to $\models$.

Definition. *$I : M \mathrel{\widetilde{\lll}_p} N$, I is a* partial monomorphism *from M to N, if I is a non-empty set of monomorphisms from finite substructures of M to finite substructures of N, satisfying the following back and forth properties:*

- *If $p \in I$ and $a \in |M|$, then there exists a $q \in I$ such that $p \subseteq q$ and $a \in dom(q)$.*
- *If $p \in I$ and $b \in |N|$, then there exists a $q \in I$ such that $p \subseteq q$ and $b \in rng(q)$.*

$M \mathrel{\widetilde{\lll}_p} N$, M is partially monomorphic *to N, if $I : M \mathrel{\widetilde{\lll}_p} N$ for some I. Partial* isomorphism, $\cong_p$, *is defined analogously.*

Theorem. *Let ϕ be a signed sentence. If $M \mathrel{\widetilde{\lll}_p} N$ and $M \models \phi$, then $N \models \phi$.*

This follows from the next lemma. For any variable assignment B into $dom(q)$, let $q \circ B$ be the variable assignment into $rng(q)$, such that $(q \circ B)(x) = q(B(x))$ for all x. Moreover, let $q \circ \Gamma$ be $\{\varphi[q \circ B]^{\odot} \mid \varphi[B]^{\odot} \in \Gamma\}$, and $rng(\Gamma)$ be $\bigcup_{\varphi[B]^{\odot} \in \Gamma} rng(B)$.

Lemma. *Suppose $I : M \mathrel{\widetilde{\lll}_p} N$. Then for all $p \in I$ and Γ such that $rng(\Gamma) \subseteq dom(p)$, if $M \models_{\alpha} \Gamma$, then $N \models p \circ \Gamma$.*

Proof: By induction on α. For $\alpha = 0$, this follows immediately from the definition of *monomorphism* and the first five conditions in the definition of $\models$. The induction steps for $\vee$ and $\neg$ are straightforward; we consider the steps for $\exists$.

- Suppose the last step in a verification tree of depth α is of the form

  $$\frac{M \models \Gamma, \varphi[A(a/x)]^{+}}{M \models \Gamma, \exists x \varphi[A]^{+}.}$$

  Now suppose $p \in I$, $rng(A) \subseteq dom(p)$ and $rng(\Gamma) \subseteq dom(p)$. There exists a $q \in I$ such that $p \subseteq q$ and $a \in dom(q)$. Hence $rng(A(a/x)) \subseteq dom(q)$ and $rng(\Gamma) \subseteq dom(q)$. Thus by the induction hypothesis

  $$N \models q \circ \Gamma, \varphi[q \circ (A(a/x))]^{+},$$

  i.e.,

  $$N \models q \circ \Gamma, \varphi[(p \circ A)(q(a)/x)]^{+}$$

  and hence by a step of existential quantification we get

  $$N \models p \circ \Gamma, \exists x \varphi[p \circ A]^{+}.$$

- Suppose the last step in a verification tree of depth α is of the form

$$\frac{\{M \models \Gamma, \varphi[A(a/x)]^-\}_{a \in |M|}}{M \models \Gamma, \exists x \varphi[A]^-.}$$

Now suppose $p \in I$, $rng(A) \subseteq dom(p)$ and $rng(\Gamma) \subseteq dom(p)$. Let b be an arbitrary element of $|N|$. Then there exists a $q \in I$ such that $p \subseteq q$ and $b \in rng(q)$. Hence $rng(A(q^{-1}(b)/x) \subseteq dom(q)$ and $rng(\Gamma) \subseteq dom(q)$. We know from the assumption that

$$M \models \Gamma, \varphi[A(q^{-1}(b)/x)]^-,$$

and therefore by the induction hypothesis that

$$N \models q \circ \Gamma, \varphi[q \circ (A(q^{-1}(b)/x))]^-,$$

i.e.,

$$N \models p \circ \Gamma, \varphi[(p \circ A)(b/x)]^-.$$

Since this holds for all b, it now follows by negative existential quantification that

$$N \models p \circ \Gamma, \exists x \varphi[p \circ A]^-.$$ □

On the other hand, it is easy to see that the existence of partial monomorphisms in both directions is *not* a *necessary* condition for semantical equivalence relative to $\models$. For complete structures, $\models^*$ coincides with $\models$, so for complete structures $\models^*$ has no greater expressive power than $\models$.

3.6 Predictive and Saturated Formulas

Perhaps the most direct way to compare $\models$ and $\models_\Box$ is through a study of predictive formulas. We recall that the formula φ is predictive iff for all M and A it satisfies the equivalence

$$M \models_\Box \varphi[A]^+ \quad \text{iff} \quad M \models \varphi[A]^+.$$

We start with a few examples of predictive formulas:

A formula is *positive* if it is built up using $\exists$, $\vee$ and the non-primitive symbols $\forall$ and $\wedge$, but contains no occurrences of $\neg$ except those of $\forall$ and $\wedge$.

Theorem. *Every positive formula is predictive.*

Proof: For a given partial model M, let M^o be its minimal completion, defined by

- $R^{M^o+} = R^{M^+}$.

- $R^{M^o-} = |M|^{n_R} - R^{M^+}$.

Since $M \models (\varphi \wedge \psi)[A]^+$ iff $M \models \varphi[A]^+$ and $M \models \psi[A]^+$, and similarly for universal quantification, we see that only the positive extensions are relevant to the truth of a positive formula. Hence it is readily seen that any positive formula φ will satisfy the equivalence

$$M^o \models \varphi[A] \quad \text{iff} \quad M \models \varphi[A].$$

But then φ must be predictive, since by definition $M \models_\Box \varphi[A]$ implies $M^o \models \varphi[A]$. □

Lemma. *Any conjunction of predictive formulas is itself predictive. If φ is predictive, then so is $\forall x\varphi$.*

Proof: We prove the first part; the second part is analogous. If $M \models_\Box (\varphi \wedge \psi)[A]$, then $M \models_\Box \varphi[A]$ and $M \models_\Box \psi[A]$. Hence if φ and ψ are predictive then also $M \models \varphi[A]$ and $M \models \psi[A]$, and thus $M \models (\varphi \wedge \psi)[A]$. □

As a trivial consequence of the theorem above we also see that every formula positively equivalent to a positive formula is predictive. This cannot be strengthened to a *characterization*, however. To see this, observe that in the proof the important property is not really that only the positive extension of any relation symbol is relevant, but rather that the positive and negative extensions are not *both* relevant. Hence any formula is predictive if the negation symbol only occurs at the atomic level, and for each relation symbol R different from identity, the negation symbol occurs in front of *all* or *none* of the atomic subformulas containing R.

But note that even this would not give a syntactical characterization of predictiveness, since also a formula like

$$\neg(x = y) \wedge (P(x) \vee \neg P(y))$$

is predictive. We leave it as an open question whether there exists a recursive set of formulas such that any formula is predictive iff it positively equivalent to a formula of this set. By the above observations, however, such a set would at least have a fairly complicated structure.

A different set of questions apply for *classical* equivalence. We have seen that in sentential logic, every formula is classically equivalent to some predictive formula. Clearly this is not so in predicate logic, since it would imply the relation $\models_\Box \leq \models^{wscl}$. We can elaborate on this observation, and for this purpose we introduce a more general notion:

Definition. *A formula* φ *is* $\models_x$ predictive *if for all* M *and* A *it satisfies the equivalence*

$$M \models_x \varphi[A]^+ \quad \textit{iff} \quad M \models \varphi[A]^+.$$

$\models\!\!=^*$ predictiveness and $\models\!\!=^*_\omega$ predictiveness represent successive weakenings of $\models_\Box$ predictiveness. We observe that not every formula is classically equivalent to some $\models\!\!=^*$ predictive formula, since this would imply the relation $\models\!\!=^* \leq \models^{wscl}$. More surprisingly, however, we shall see that neither is every formula classically equivalent to some $\models\!\!=^*_\omega$ predictive formula. To reach this result, we introduce an inference-theoretical counterpart to the notion of predictive formulas:

Definition. *The formula* φ *is* saturated *if* $\models_2 \psi \supset \varphi$ *implies* $\models_3 \psi \supset \varphi$ *for every formula* ψ.

Theorem. *A formula is saturated iff it is* $\models\!\!=^*_\omega$ *predictive.*

Proof: Let φ be $\models\!\!=^*_\omega$ predictive, and suppose $\models_2 \psi \supset \varphi$ and $M \models_3 \psi[A]$. We conclude consecutively that $M \models\!\!=_\omega \psi[A]$ and $M \models\!\!=_\omega \varphi[A]$. Since φ is $\models\!\!=^*_\omega$ predictive, $M \models \varphi[A]$. Since M and A were arbitrary, we have shown $\models_3 \psi \supset \varphi$.

Next suppose $M \models\!\!=_\omega \varphi[A]$ and φ is saturated. From the first assumption it follows that $M \models \psi[A]$ for a formula ψ such that $\models_2 \psi \supset \varphi$. But then $\models_3 \psi \supset \varphi$ since φ is saturated. Hence $M \models \varphi[A]^+$. □

The next theorem states that saturatedness is also a fairly strong property. We show that the negation of a sentence expressing transitivity is not classically equivalent to any saturated formula. We need the following lemma.

Lemma. *If* φ *is saturated, then so is* $\forall x\varphi$.

Proof: Suppose φ is saturated and $\models_2 \psi \supset \forall x\varphi$. We get the following procession of inferences: $\models_2 \exists x\psi \supset \forall x\varphi$, $\models_2 \exists x\psi \supset \varphi$, $\models_3 \exists x\psi \supset \varphi$, $\models_3 \exists x\psi \supset \forall x\varphi$, $\models_3 \psi \supset \forall x\varphi$. □

Theorem. *Not every formula is classically equivalent to a saturated formula. Not every sentence is classically equivalent to a saturated sentence.*

Proof: We show that the sentence

$$\exists x_1 \exists x_2 \exists x_3 (R(x_1, x_1) \wedge R(x_2, x_3) \wedge \neg R(x_1, x_3))$$

is not classically equivalent to any saturated formula. For each $n > 2$ let ϕ_n be the sentence

$$\exists x_1 \ldots \exists x_n (R(x_1, x_2) \wedge \ldots \wedge R(x_{n-1}, x_n) \wedge \neg R(x_1, x_n)).$$

Similarly, for each $n > 2$ let $\langle a_1, \ldots, a_n \rangle$ be a *non-transitive chain* (*ntc*) in M iff $\langle a_i, a_{i+1} \rangle \in R^{M^+}$ for all i: $1 \leq i < n$, and $\langle a_1, a_n \rangle \in R^{M^-}$. Clearly M contains an *ntc* of length n iff $M \models \phi_n$, and M contains some *ntc* iff $M \models \bigvee\!\!\bigvee_{2<n<\omega} \phi_n$.

First observe that $\models_2 \varphi_n \supset \varphi_3$ for any $n \geq 3$. To see this, suppose M is complete and contains an *ntc* of length greater than 3. It is sufficient to show that M in this case must contain a shorter *ntc* as well. So suppose $\langle a_1, a_2 \rangle, \ldots, \langle a_{n-1}, a_n \rangle \in R^{M^+}$ and $\langle a_1, a_n \rangle \in R^{M^-}$. If $\langle a_{n-2}, a_n \rangle \in R^{M-}$ then there is an *ntc* of length 3, and we are done. Otherwise $\langle a_{n-2}, a_n \rangle \in R^{M^+}$, and there is an *ntc* of length $n - 1$.

Next observe that if the model M contains no *ntc*, then there is also a completion N that does not contain an *ntc*:

- $\langle a, b \rangle \in R^{N^+}$ iff for some $n \geq 2$ there exist $a_1, \ldots, a_n$ where $a = a_1$, $b = a_n$ and $\langle a_i, a_{i+1} \rangle \in R^{M^+}$ for all i: $1 \leq i < n$.
- $R^{N^-} = |M|^2 - R^{N^+}$.

This definition guarantees that R^{N^+} is transitive, hence N does not contain an *ntc*. Moreover, since M itself does not contain an *ntc*, $R^{M^-} \cap R^{N^+} = \emptyset$. Since also $R^{M^+} \subseteq R^{N^+}$ it follows that N is a completion of M.

For a reductio ad absurdum we now suppose that some formula is saturated and classically equivalent to ϕ_3. Since ϕ_3 is a sentence, also the universal closure of this formula is classically equivalent to ϕ_3. By the above lemma this universal closure is saturated as well. Hence there is a *sentence* σ which is both saturated and classically equivalent to ϕ_3. We first show that such a σ would have to satisfy $\models_3 \sigma \equiv \bigvee\!\!\bigvee_{2<n<\omega} \phi_n$:

We have already seen that $\models_2 \phi_n \supset \phi_3$ for every $n \geq 3$. Therefore, since σ is classically equivalent to ϕ_3, also $\models_2 \phi_n \supset \sigma$ and hence $\models_3 \phi_n \supset \sigma$ since σ is saturated. Since this holds for all n, $\models_3 \bigvee\!\!\bigvee_{2<n<\omega} \phi_n \supset \sigma$. For the other direction, suppose $M \models \sigma$ for a partial model M. By persistence $M \models_\Box \sigma$ and thus $M \models_\Box \phi_3$. Hence every completion of M contains an *ntc*, so M itself contains an *ntc* by the argument above, wherefore $M \models \bigvee\!\!\bigvee_{2<n<\omega} \phi_n$.

We have now shown that $M \models \sigma$ iff M contains an *ntc*. From this we conclude that σ is preserved under extension of domain. If $N \uparrow D$ contains an *ntc*, then certainly so does N. Hence σ is positively equivalent to an existential sentence τ.

Finally we show that this is impossible; there is no existential sentence that is true in a model iff this model contains an *ntc*. τ is of the form $\exists x_1 \ldots \exists x_n \tau_0$ for some n and quantifier free formula τ_0 with at most $x_1, \ldots, x_n$ free. Now let M be a model where

- $R^{M^+} = \{\langle a_1, a_2 \rangle, \ldots, \langle a_n, a_{n+1} \rangle\}$.

- $R^{M^-} = \{\langle a_1, a_{n+1}\rangle\}$

and all the $a_1, \ldots, a_{n+1}$ are distinct. Then M contains an *ntc*, so $M \models \tau$. Thus $M \models \tau_0[A]$ for some A. Since there are at most n free variables in τ_0 and the *ntc* in M has $n+1$ elements, there must be an $i : 1 \leq i \leq n+1$ such that $A(y) \neq a_i$ for all the free variables y of τ_0. We derive from M the model M_0 by deleting pairs $\langle a_i, a_j\rangle$ or $\langle a_j, a_i\rangle$, for the i above, from R^{M^+}. This does not affect the evaluation of τ_0 with respect to A, hence $M_0 \models \tau_0[A]$, and hence $M_0 \models \tau$. But this is impossible since M_0 contains no *ntc*. □

We have seen that the claim

$$\forall\varphi\exists\varphi'\forall\psi(\models_2 \varphi \equiv \varphi' \quad \& \quad (\models_2 \psi \supset \varphi' \quad \rightarrow \quad \models_3 \psi \supset \varphi'))$$

does not hold. However, we can weaken this by switching the the quantifiers:

$$\forall\varphi\forall\psi\exists\varphi'(\models_2 \varphi \equiv \varphi' \quad \& \quad (\models_2 \psi \supset \varphi' \quad \rightarrow \quad \models_3 \psi \supset \varphi')).$$

This claim is true; we can even strengthen it to the following:

$$\forall\varphi\forall\psi\exists\varphi'(\models_3 \neg\varphi \equiv \neg\varphi' \quad \& \quad (\models_2 \psi \supset \varphi' \quad \rightarrow \quad \models_3 \psi \supset \varphi')).$$

We say that φ' is saturated relative to ψ if it satisfies the material implication $\models_2 \psi \supset \varphi' \ \rightarrow\ \models_3 \psi \supset \varphi'$, i.e., if either $\models_2 \psi \supset \varphi'$ does *not* hold, or $\models_3 \psi \supset \varphi'$ *does* hold. The last claim above expresses the relative saturation theorem:

Theorem. *For every two formulas φ and ψ there exists a formula χ such that $\models_3 \neg\varphi \equiv \neg\chi$, and $\models_2 \psi \supset \chi$ implies $\models_3 \psi \supset \chi$.*

We also list a slightly different form of the theorem. The two versions are obviously equivalent:

Theorem. *For every two formulas φ and ψ there exists a formula χ such that $\models_3 \neg\varphi \equiv \neg\chi$, and $\models_2 \psi \supset \varphi$ implies $\models_3 \psi \supset \chi$.*

Proof: If $\not\models_2 \psi \supset \varphi$, let χ be φ. Now suppose $\models_2 \psi \supset \varphi$, and for arbitrary M and A suppose $M \models \psi[A]$. Now $M \models\!\!=_\omega \psi[A]$ and thus $M \models\!\!=_\omega \varphi[A]$ since $\models\!\!=_\omega$ is closed under classical consequence. Hence $M \models \varphi_{M,A}[A]$ for a formula $\varphi_{M,A}$ such that $\models_3 \neg\varphi \equiv \neg\varphi_{M,A}$. Let Π_φ be the set

$$\{\varphi' \mid \models_3 \neg\varphi \equiv \neg\varphi'\}.$$

We have shown that for every M and A such that $M \models \psi[A]$, we have $M \models \varphi'[A]$ for some $\varphi' \in \Pi_\varphi$. In other words

$$\psi \models_3 \Pi_\varphi.$$

By compactness $\psi \models_3 \Pi_0$ for some finite subset Π_0 of Π_φ. Hence $\models_3 \psi \supset \bigvee\!\!\!\bigvee \Pi_0$. Clearly also $\models_3 \neg\varphi \equiv \neg \bigvee\!\!\!\bigvee \Pi_0$, so $\bigvee\!\!\!\bigvee \Pi_0$ is a relative saturant. □

As an immediate consequence of the relative saturation theorem, we get the "cut-and-glue" theorem:

Theorem.

(i) *For every two classically equivalent formulas there is a third formula, positively equivalent to the first and negatively equivalent to the second.*

(ii) *For every two classically equivalent sentences there is a third sentence, positively equivalent to the first and negatively equivalent to the second.*

Proof: (i) This follows from the relative saturation theorem by the same argument that was used for **1**.

(ii) Suppose φ and ψ are sentences, and $\models_2 \varphi \equiv \psi$. Then by *(i)* there exists a formula χ such that

$$\models_3 \varphi \equiv \chi \quad \text{and} \quad \models_3 \neg\psi \equiv \neg\chi.$$

Since φ and ψ are sentences, it then follows that

$$\models_3 \varphi \equiv \chi' \quad \text{and} \quad \models_3 \neg\psi \equiv \neg\chi'$$

where χ' is the existential closure of χ. □

The relative saturation theorem appears as a joint corollary to compactness and several results about $\models_\omega$. Hence the proof relies on the extensive apparatus developed in this chapter. One might speculate on how much of that apparatus is really needed to prove the relative saturation theorem, and whether a simpler shortcut might exist. The next result is an indication that at least a certain level of complexity will be involved in any proof:

Theorem. *There is no recursive function that for every two formulas φ and ψ will yield a formula χ, classically equivalent to ψ, and saturated relative to φ.*

Proof: Suppose otherwise. Let $sat_\psi(\varphi)$ be the result of applying this function, and consider the case where ψ is $\top$. For any formula φ we get

(i) $\models_2 sat_\top(\varphi) \equiv \varphi$ and

(ii) $\models_2 sat_\top(\varphi) \quad \rightarrow \quad \models_3 sat_\top(\varphi)$.

Hence $\models_3 sat_\top(\varphi)$ iff $\models_2 \varphi$. But this is impossible, since the set of of valid formulas with respect to $\models_3$ is recursive. □

4

Extending the Language

In the previous chapter we carried out a systematic comparison between various alternative truth definitions for the sentences of a standard first order language, relative to the partial structures. The focus of these investigations was on the pair $\langle \mathcal{L}, \models \rangle$, i.e., the syntax of $\mathcal{L}$ under the strong Kleene truth definition. In this chapter we turn to a systematic comparison between $\langle \mathcal{L}, \models \rangle$ and possible *extensions.* Hence we shall shift the perspective but keep the focus.

We shall call $\langle \mathcal{L}, \models \rangle$ a *language.* Hence a language not only has a syntactic component, but also identifies a corresponding truth definition. By the *languages* $\mathcal{L}$ and $\mathbf{l}$ we shall mean $\langle \mathcal{L}, \models \rangle$ and $\langle \mathbf{l}, \models \rangle$.

The truth functional completeness result for $\mathbf{l}$ is a result about maximal expressive power. Define a language $\mathbf{l}^1$ to have at least the expressive power of $\mathbf{l}^2$ iff for every ρ and $\mathbf{l}^2[\rho]$ sentence φ_2 there is an $\mathbf{l}^1[\rho]$ sentence φ_1 such that for all ρ models v, $v \models_{\mathbf{l}^1} \varphi_1^{\odot}$ iff $v \models_{\mathbf{l}^2} \varphi_2^{\odot}$. If $\mathbf{l}^1$ is a language to describe propositional models, then every $\mathbf{l}^1[\rho]$ sentence defines a function from the set of ρ models into $\{1, 0, \wr, \times\}$. Hence from the truth functional completeness result it follows that $\mathbf{l}$ under strong Kleene has maximal expressive power among languages that are coherent, determinable and persistent.

When we turn to predicate logic, it is natural to search for a coherent, determinable and persistent language that in a similar way has maximal expressive power with respect to the partial models we have introduced for predicate logic.

However, such a search would take us well into higher order predicate logic, and we shall not pursue it here. Instead, we shall add as conditions some well known properties of first order predicate logic, and show a corresponding maximality result for $\mathcal{L}$ under strong Kleene, among languages that are coherent, determinable and persistent. The conditions we add are compactness and the Löwenheim property. Hence we shall prove a counterpart to Lindström's characterization theorem for classical model theory, cf. Lindström (1966, 1969).

In a simple form, this theorem states that every sentence of a (classically) compact extension of $\langle \mathcal{L}, \models_2 \rangle$ that satisfies a certain Skolem–Löwenheim property is classically equivalent to an $\mathcal{L}$ sentence.

Both mathematical practice, as well as language use in general, often call for more powerful language tools than those of first order logic. Hence the Lindström theorem for partial model theory does in no way identify $\langle \mathcal{L}, \models \rangle$ as *the* most powerful language in which to do partial logic, and there is certainly a call for additional research on the expressive powers of persistent languages. On the other hand, the result *does* tell us that if coherence, determinability and persistence are conditions that should be observed when we generalize from classical to partial logic, then there are very good reasons to accept $\mathcal{L}$ under strong Kleene as the natural counterpart to standard first order languages. This issue is brought up in the next section. There is a sense in which we can say *less* in $\langle \mathcal{L}, \models \rangle$ about the partial models than we can say in standard first order languages about classical models. $\mathcal{L}_\sim$ is $\mathcal{L}$ with exclusion negation added; when we pass from $\mathcal{L}$ to $\mathcal{L}_\sim$ more things become expressible and the balance is regained. Of course this also buys us non-persistence as a side effect, but *how pervasive* is this side effect? Does the step from $\mathcal{L}$ to $\mathcal{L}_\sim$ *only* provide us with non-persistent propositions, or will there be persistent sentences of $\mathcal{L}_\sim$ that are not strongly equivalent to any $\mathcal{L}$ sentence? We have seen that $\mathbf{l}$ corresponds to the persistent fragment of $\mathbf{l}_\sim$, but the complex interaction in predicate logic between quantifiers and propositional connectives makes it necessary to raise the analogous question again; there is no easy reduction from the predicate logic to the propositional logic case.

As a matter of fact, however, it *does* turn out that in a similar way, every persistent sentence of $\mathcal{L}_\sim$ is strongly equivalent to a sentence of $\mathcal{L}$. Put differently, there is no persistent language with expressive powers strictly between those of $\mathcal{L}$ and $\mathcal{L}_\sim$. The Lindström theorem is a generalization of this result. If we want to strengthen the language $\langle \mathcal{L}, \models \rangle$ in however small and innocent a way, while retaining coherence, determinability and persistence, then we will have to take the plunge outside the familiar domain where compactness and Skolem–Löwenheim properties hold.

4.1 Semantical Equivalence for $\mathcal{L}_\sim$

The condition of persistence posits a limitation on the expressive power of a language. We have seen that two structures M and N are semantically equivalent, relative to $\mathcal{L}$, iff $M \mathrel{\widetilde{\ll}_\omega} N$ and $N \mathrel{\widetilde{\ll}_\omega} M$. This conjunctive condition is weaker than $M \cong_\omega N$, which corresponds to standard elementary equivalence. We shall see that for the language $\mathcal{L}_\sim$, semantical equivalence between partial structures corresponds to $\cong_\omega$. In addition to its inherent interest and the motivation it provides for the subsequent quest for persistence characterization results, this result will play a crucial part in the proof of the Lindström theorem.

The formulas of $\mathcal{L}_\sim$ are built up from the atomic formulas using $\exists, \vee, \neg, \sim$; we recall the semantic clauses for *exclusion negation*:

$$M \models \sim\varphi[A]^+ \quad \text{iff} \quad M \not\models \varphi[A]^+$$

$$M \models \sim\varphi[A]^- \quad \text{iff} \quad M \models \varphi[A]^+.$$

In order to prove the characterization of semantical equivalence for $\mathcal{L}_\sim$, we shall again make use of the correspondence between generalized and classical structures. Following exactly the strategy in chapter 2, for any generalized structure M for similarity type ρ we define a classical structure M^* for the similarity type ρ^* with two relation symbols R^+ and R^- for each relation symbol R in ρ. The positive and negative translations of $\mathcal{L}[\rho]$ formulas into $\mathcal{L}_{\wedge,\forall}[\rho^*]$ formulas, are extended to $\mathcal{L}_\sim[\rho]$ formulas in general by adding the clauses

$$(\sim\varphi)^* \quad = \quad \neg\varphi^*.$$

$$(\sim\varphi)_* \quad = \quad \varphi^*.$$

For classical structures the two negation symbols are equivalent. Hence '$\sim$' is translated to '$\neg$', and the result of a translation will still be an $\mathcal{L}_{\wedge,\forall}[\rho^*]$ formula.

Lemma. *For any $\mathcal{L}_\sim$ formula φ and generalized model M, $M \models \varphi[A]^+$ iff $M^* \models \varphi^*[A]^+$, and $M \models \varphi[A]^-$ iff $M^* \models \varphi_*[A]^+$.*

Proof: See the corresponding proofs for $\mathcal{L}$ and $\mathbf{l}_{\sim,\star,\diamond}$. □

We can now extend the *inverse* $^{-*}$ to be defined for *all* $\mathcal{L}_{\wedge,\forall}[\rho^*]$ formulas, by adding the clause

$$(\neg\varphi)^{-*} = \sim\varphi^{-*}$$

when φ is not an atomic identity formula. Hence * maps the set of $\mathcal{L}_\sim[\rho]$ formulas *onto* the set of $\mathcal{L}_{\wedge,\forall}[\rho^*]$ formulas. Moreover, since φ is a sentence iff φ^* is a sentence and iff φ_* is a sentence, we obtain for all partial models M and N:

$$\forall\phi \in \mathcal{L}_\sim[\rho](M \models \phi \leftrightarrow N \models \phi) \quad \text{iff} \quad \forall\tau \in \mathcal{L}_{\wedge,\forall}[\rho^*](M^* \models \tau \leftrightarrow N^* \models \tau).$$

Here, ϕ ranges over signed sentences and τ ranges over unsigned (positively signed) sentences.

In classical first order logic, semantical equivalence between structures, relative to $\mathcal{L}$ or $\mathcal{L}_{\wedge,\forall}$, corresponds to the existence of an ω-partial isomorphism. This is a well-known result, but can also be deduced as a special case of the corresponding characterization of semantical equivalence relative to partial structures, which we already proved in chapter 2. This is because the notions of ω-partial *homo*morphisms and *iso*morphisms coincide for complete structures. Hence

$$M^* \cong_\omega N^* \quad \text{iff} \quad \forall \tau \in \mathcal{L}_{\wedge,\forall}[\rho^*](M^* \models \tau \leftrightarrow N^* \models \tau).$$

Since also

$$M \cong_\omega N \quad \text{iff} \quad M^* \cong_\omega N^*,$$

it follows that

$$M \cong_\omega N \quad \text{iff} \quad \forall \phi \in \mathcal{L}_\sim[\rho](M \models \phi \leftrightarrow N \models \phi).$$

We have proved:

Theorem. *Let M and N be partial ρ models. $M \cong_\omega N$ iff for all signed $\mathcal{L}_\sim[\rho]$ sentences ϕ, $M \models \phi$ iff $N \models \phi$.*

4.2 Maximality of $\mathcal{L}_\sim$

This section contains the main step in the proof of the Lindström theorem. The main lemma of the section states the maximality of $\mathcal{L}_\sim$ among coherent and determinable languages satisfying two properties familiar from standard first order logic.

Lindström's original theorem, as well as our version of it, are results about arbitrary extensions of a language. Before we start the discussions of this section we should therefore make clear how we understand these notions. $\mathcal{L}[\rho]$ is the set of $\mathcal{L}$ sentences for similarity type ρ. Similarly for $\mathcal{L}_\sim$. In general, if $\mathcal{L}^*$ is a language, then for every similarity type ρ, $\mathcal{L}^*$ identifies a set $\mathcal{L}^*[\rho]$ of ρ-sentences. For each ρ, $\mathcal{L}^*$ identifies a semantical relation $\models_{\mathcal{L}^*}$ between ρ models and *signed* $\mathcal{L}^*[\rho]$ sentences. We lay down the following necessary conditions, which are adapted from Ebbinghaus (1983).

($C1$) If $\rho \subseteq \rho'$, then $\mathcal{L}^*[\rho] \subseteq \mathcal{L}^*[\rho']$.

($C2$) If $M \models_{\mathcal{L}^*} \phi$ and $M \cong N$, then $N \models_{\mathcal{L}^*} \phi$.

($C3$) If M is a ρ' structure, $\rho \subseteq \rho'$ and $\phi \in \mathcal{L}^*[\rho]$, then $M \models_{\mathcal{L}^*} \phi$ iff $M \upharpoonright \rho \models_{\mathcal{L}^*} \phi$.

(*C*4) Let $\sigma : \rho \to \rho'$ be a *renaming*, i.e., a bijection preserving arities. For any ρ structure M, let M' be the ρ' structure with the same domain as M, where $\sigma(R)$ is interpreted as R is in M. Then for each $\varphi \in \mathcal{L}^*[\rho]$ there is a $\varphi' \in \mathcal{L}^*[\rho']$ such that for all M for similarity type ρ, $M \models_{\mathcal{L}^*} \varphi^{\odot}$ iff $M' \models_{\mathcal{L}^*} \varphi'^{\odot}$.

We use all of these assumptions in the proof of the Lindström theorem. In addition, $\mathcal{L}^*$ is *persistent* if for any signed $\mathcal{L}^*$ sentence ϕ, $M \models_{\mathcal{L}^*} \phi$ and $M \ll N$ together imply $N \models_{\mathcal{L}^*} \phi$. Similarly, $\mathcal{L}^*$ is *coherent* if $M \models_{\mathcal{L}^*} \varphi^+$ implies $M \not\models_{\mathcal{L}^*} \varphi^-$ for all structures M and $\mathcal{L}^*$ sentences φ. $\mathcal{L}^*$ is *determinable* if for any complete ρ model M and any $\varphi \in \mathcal{L}^*[\rho]$, $M \models_{\mathcal{L}^*} \varphi^+$ or $M \models_{\mathcal{L}^*} \varphi^-$.

By an *extension* of a language $\mathcal{L}^*$ we mean a language $\mathcal{L}^*$ such that $\mathcal{L}^*[\rho]$ is a subset of $\mathcal{L}^*[\rho]$ for all ρ, and for all $\phi \in \mathcal{L}^*$ we have $M \models_{\mathcal{L}^*} \phi$ iff $M \models_{\mathcal{L}^*} \phi$.

Definition. *$\mathcal{L}^*$ has the* Skolem property *if for every ρ and every ρ model M there exists a countable neighborhood assignment ξ such that for any non-empty subset E of $|M|$ closed under ξ and for all signed $\mathcal{L}^*[\rho]$ sentences ϕ, $M \models_{\mathcal{L}^*} \phi$ iff $M \upharpoonright E \models_{\mathcal{L}^*} \phi$.*

Definition. *$\mathcal{L}^*$ has the* Löwenheim property *if for every ρ and every ρ model M there exists a countable model M_0 such that for all signed $\mathcal{L}^*[\rho]$ sentences ϕ, $M \models_{\mathcal{L}^*} \phi$ iff $M_0 \models_{\mathcal{L}^*} \phi$.*

Clearly the Skolem property implies the Löwenheim property, since for any countable neighborhood assignment on a model there is a non-empty, countable subset of the domain of the model that is closed under the given neighborhood assignment.

Let Γ and Δ be sets of signed $\mathcal{L}^*[\rho]$ sentences. $\Gamma \models_{\mathcal{L}^*} \Delta$ holds iff for every ρ model M such that $M \models_{\mathcal{L}^*} \phi$ for all $\phi \in \Gamma$, also $M \models_{\mathcal{L}^*} \phi'$ for some $\phi' \in \Delta$. By C3, this consequence relation is independent of the similarity type ρ.

Definition. *$\mathcal{L}^*$ is* countably compact, *if for every ρ and every two countable sets Γ and Δ of signed $\mathcal{L}^*[\rho]$ sentences, if $\Gamma \models_{\mathcal{L}^*} \Delta$, then also $\Gamma_0 \models_{\mathcal{L}^*} \Delta_0$ for some finite subsets Γ_0 and Δ_0 of Γ and Δ respectively.*

When $\mathcal{L}^*$ is an extension of $\mathcal{L}$, and there is no possibility of confusion, we will sometimes write $M \models \phi$ or $M \models_3 \phi$ for $M \models_{\mathcal{L}^*} \phi$, and $\Gamma \models_3 \Delta$ for $\Gamma \models_{\mathcal{L}^*} \Delta$. Some minor modifications of arguments from the literature will take us to our first Lindström type characterization result, which states a certain maximality of $\mathcal{L}_\sim$ up to positive equivalence. The strategy used is a combination of the strategies found in Barwise (1974) and Flum (1983).

In the following proofs we shall refer to the sentence $\sim\psi$ obtained from an $\mathcal{L}^*$ sentence ψ. This sentence is true iff ψ is not true, and false iff ψ is true. $\sim\psi$ itself may not be an $\mathcal{L}^*$ sentence, but the notation is harmless since a corresponding extension of $\mathcal{L}^*$ is clearly possible. When we apply compactness, the Löwenheim property, etc., there will be no assumption that the language $\mathcal{L}^*$ is itself closed under exclusion negation. In a similar way we shall sometimes refer to the sentence $(\varphi \wedge \psi)$ for arbitrary $\mathcal{L}^*$ sentences φ and ψ, without thereby assuming that this itself is an $\mathcal{L}^*$ sentence.

We prove that if ψ is a sentence of a countably compact extension of $\mathcal{L}_\sim$ that satisfies the Löwenheim property, then ψ is positively equivalent to an $\mathcal{L}_\sim$ sentence. The proof is by contraposition. If ψ is a sentence of such an extension, and not positively equivalent to any $\mathcal{L}_\sim$ sentence, then neither is $\sim\psi$ positively equivalent to any $\mathcal{L}_\sim$ sentence. Next we strengthen ψ and $\sim\psi$ to new sentences, neither of which are equivalent to any $\mathcal{L}_\sim$ sentence (and therefore have models), by adding the *same* $\mathcal{L}_\sim$ sentences as conjuncts to both. These additions are done iteratively, with the added $\mathcal{L}_\sim$ sentences converging to a maximal consistent set of $\mathcal{L}_\sim$ sentences. (A *consistent* set of sentences is a set of sentences that has a model.) By compactness we obtain a maximal consistent set T of $\mathcal{L}_\sim$ sentences such that $T \cup \{\psi\}$ and $T \cup \{\sim\psi\}$ have a model each. Since both are models of the same maximal consistent set of $\mathcal{L}_\sim$ sentences, the two models are semantically equivalent relative to $\mathcal{L}_\sim$, and hence ω-partially isomorphic. The second part of the proof shows (provided $\mathcal{L}^*$ also satisfies the Löwenheim property) that in this case there also exist two *isomorphic* models for T, exactly one of which makes ψ true. But this violates our assumption that $\mathcal{L}^*$ does not distinguish between isomorphic structures, and hence the theorem is proved. We start with the following lemma:

Lemma. *Let $\mathcal{L}^*$ be an extension of $\mathcal{L}_\sim$. Moreover, let ψ be an $\mathcal{L}^*[\rho]$ sentence, and let both χ and φ be $\mathcal{L}_\sim[\rho]$ sentences. Suppose that both*

(i) *either $(\psi \wedge \chi \wedge \varphi)$ or $(\sim\psi \wedge \chi \wedge \varphi)$ is positively equivalent to an $\mathcal{L}_\sim[\rho]$ sentence, and*

(ii) *either $(\psi \wedge \chi \wedge \sim\varphi)$ or $(\sim\psi \wedge \chi \wedge \sim\varphi)$ is positively equivalent to an $\mathcal{L}_\sim[\rho]$ sentence.*

Then either $(\psi \wedge \chi)$ or $(\sim\psi \wedge \chi)$ is positively equivalent to an $\mathcal{L}_\sim[\rho]$ sentence.

Proof: Assume the antecedents. Then there exist $\mathcal{L}_\sim$ sentences μ_0 and μ_1 such that either

(i) $\models_3 (\psi \wedge \chi \wedge \varphi) \equiv \mu_0$ and $\models_3 (\psi \wedge \chi \wedge \sim\varphi) \equiv \mu_1$, or

(ii) $\models_3 (\psi \wedge \chi \wedge \varphi) \equiv \mu_0$ and $\models_3 (\sim\psi \wedge \chi \wedge \sim\varphi) \equiv \mu_1$, or

(iii) $\models_3 (\sim\psi \wedge \chi \wedge \varphi) \equiv \mu_0$ and $\models_3 (\psi \wedge \chi \wedge \sim\varphi) \equiv \mu_1$, or

(iv) $\models_3 (\sim\psi \wedge \chi \wedge \varphi) \equiv \mu_0$ and $\models_3 (\sim\psi \wedge \chi \wedge \sim\varphi) \equiv \mu_1$.

Now observe that

(i) yields $\models_3 (\psi \wedge \chi) \equiv (\mu_0 \vee \mu_1)$

(ii) yields $\models_3 (\psi \wedge \chi) \equiv (\mu_0 \vee (\sim\mu_1 \wedge \chi \wedge \sim\varphi))$

(iii) yields $\models_3 (\sim\psi \wedge \chi) \equiv (\mu_0 \vee (\sim\mu_1 \wedge \chi \wedge \sim\varphi))$

(iv) yields $\models_3 (\sim\psi \wedge \chi) \equiv (\mu_0 \vee \mu_1)$.

Hence the conclusion follows. □

Lemma. *Let $\mathcal{L}^*$ be a countably compact extension of $\mathcal{L}_\sim$, and suppose ψ is an $\mathcal{L}^*[\rho]$ sentence that is not positively equivalent to any $\mathcal{L}_\sim[\rho]$ sentence. Then there exists a maximal consistent set T of $\mathcal{L}_\sim[\rho]$ sentences such that both $T \cup \{\psi\}$ and $T \cup \{\sim\psi\}$ have models.*

Proof: Since ψ is not positively equivalent to any $\mathcal{L}_\sim[\rho]$ sentence, neither is $\sim\psi$ positively equivalent to any $\mathcal{L}_\sim[\rho]$ sentence. Now let $\langle\varphi_n\rangle_{n<\omega}$ be an enumeration of the $\mathcal{L}_\sim[\rho]$ sentences. By the previous lemma there exists a sequence $\langle\chi_n\rangle_{n<\omega}$ such that each χ_n is either φ_n or $\sim\varphi_n$, and such that for any m neither $(\psi \wedge \bigwedge_{n\leq m}\chi_n)$ nor $(\sim\psi \wedge \bigwedge_{n\leq m}\chi_n)$ is equivalent to any $\mathcal{L}_\sim$ sentence. Hence for all m both $\{\psi\} \cup \{\chi_n\}_{n\leq m}$ and $\{\sim\psi\} \cup \{\chi_n\}_{n\leq m}$ have a model, i.e., neither

$$\{\psi\} \cup \{\chi_n\}_{n\leq m} \models_3$$

nor

$$\{\chi_n\}_{n\leq m} \models_3 \psi$$

holds. We have assumed that $\mathcal{L}^*$ is countably compact. Hence neither does

$$\{\psi\} \cup \{\chi_n\}_{n<\omega} \models_3$$

or

$$\{\chi_n\}_{n<\omega} \models_3 \psi$$

hold. In other words, both $\{\chi_n\}_{n<\omega} \cup \{\psi\}$ and $\{\chi_n\}_{n<\omega} \cup \{\sim\psi\}$ have models. From the way the set is defined, we see that $\{\chi_n\}_{n<\omega}$ must be a maximal consistent set of $\mathcal{L}_\sim$ sentences, so the argument is complete. □

The following lemma is also needed. We recall the definitions of 'isomorphism' ($\cong$), 'partial isomorphism' ($\cong_p$) and 'ω-partial isomorphism' ($\cong_\omega$) from chapters 2 and 3.

Lemma.

(i) *If* $|M|$ *and* $|N|$ *are countable and* $M \cong_p N$, *then* $M \cong N$.

(ii) *If* $|M|$ *and* $|N|$ *are finite and* $M \cong_\omega N$, *then* $M \cong N$.

Proof: These results follow by simple, familiar arguments that do not differ from the corresponding proofs for classical structures. □

We are now ready to state and prove the first Lindström type characterization result.

Lemma. *Let* $\mathcal{L}^*$ *be a countably compact extension of* $\mathcal{L}_\sim$ *that satisfies the Löwenheim property. Then every* $\mathcal{L}^*[\rho]$ *sentence is positively equivalent to an* $\mathcal{L}_\sim[\rho]$ *sentence.*

Proof: Assume the antecedent, and suppose (for a reductio ad absurdum) that the $\mathcal{L}^*[\rho]$ sentence ψ is not positively equivalent to any $\mathcal{L}[\rho]$ sentence. By the previous lemma there exists a maximal consistent set T of $\mathcal{L}_\sim[\rho]$ sentences such that $\{\psi\} \cup T$ has a model $N^\bullet$ and $\{\sim\psi\} \cup T$ has a model N°. Let ρ' be a copy of the similarity type ρ, disjoint from ρ, and let $N^{\circ'}$ be the ρ' model with domain equal to the domain of N°, and where each relation symbol R' is interpreted as its counterpart R is in N°. Moreover, let T' be the set obtained from T by substituting relation symbols R' for their counterparts R, and similarly let ψ' be the sentence given by $(C4)$. Then $N^{\circ'}$ is a model for $\{\sim\psi'\} \cup T'$. By the Löwenheim property, both $\{\psi\} \cup T$ and $\{\sim\psi'\} \cup T'$ have countable models. Since T is a maximal consistent set, these models are either both countably infinite, or they are of the same finite cardinality. By $(C2)$ they have a model each where the domains are the same. By $(C3)$ $\{\psi, \sim\psi'\} \cup T \cup T'$ has a $\rho \cup \rho'$ model M.

We first observe that M cannot be finite. Let $(M \uparrow \rho')^{-'}$ be the model obtained from $M \uparrow \rho'$ by the renaming from ρ' to ρ. Now both $M \uparrow \rho$ and $(M \uparrow \rho')^{-'}$ are models of T. Since T is a maximal consistent set, we have

$$(M \uparrow \rho) \cong_\omega (M \uparrow \rho')^{-'}.$$

If $|M|$ is finite, this implies.

$$(M \uparrow \rho) \cong (M \uparrow \rho')^{-'}.$$

Since one of these models makes ψ true and the other not, this would contradict the isomorphism condition $(C2)$ on $\mathcal{L}^*$.

Hence we can assume that $|M|$ is countably infinite. The strategy for the rest of the proof is the following. We encode a partial isomorphism between $N \uparrow \rho$ and $(N \uparrow \rho')^{-'}$ for $\rho \cup \rho'$ structures N, using an infinite set

P of sentences in an expansion of the similarity type $\rho \cup \rho'$. For every finite subset P_0 of P, M can be expanded to a model of P_0. By countable compactness, $\{\psi, \sim\psi'\} \cup P$ has a model, and a hence *countable* model N. $N \upharpoonright \rho$ will now be isomorphic to $(N \upharpoonright \rho')^{-'}$. But this contradicts (C_2), since these two models will not agree on ψ.

We start by encoding the partial isomorphism. The most direct encoding requires an infinite similarity type $\{F_n\}_{n<\omega}$ where for each n the truth of $F_n(x_1, \ldots, x_n, y_1, \ldots, y_n)[A]$ implies the existence of an isomorphism with $A[\{x_1, \ldots, x_n\}]$ and $A[\{y_1, \ldots, y_n\}]$ as domain and range. Up till now we have only considered finite similarity types, however, and for the proof it is *not* necessary to assume that $\mathcal{L}^*$ is countably compact or has the Löwenheim property with respect to *infinite* similarity types. In the following strategy, which is adapted from Barwise (1974), we define complex expressions $\{E_n\}_{n<\omega}$ in a finite similarity type, that mimic the primitive symbols $\{F_n\}_{n<\omega}$. Let ϱ be the similarity type

$$\{< \text{(binary)},\ p \text{ (ternary)},\ E \text{ (ternary)}\},$$

and for each $n < \omega$ let $N_n(w)$ be the formula

$$\exists w_1 \ldots \exists w_n ((\bigwedge_{i<j\leq n} \neg(w_i = w_j)) \wedge \forall z (z < w \equiv (z = w_1 \vee \ldots \vee z = w_n)))$$

So $N_n(x)$ says that there are exactly n elements $<$-less than x. Furthermore, let $E_n(x_1, \ldots, x_n, y_1, \ldots, y_n)$ be the formula

$$\begin{aligned}\exists w \exists v_1 \ldots \exists v_{n-1} \exists u_1 \ldots \exists u_{n-1} (N_n(w) \wedge \\ p(x_1, x_2, v_1) \wedge p(v_1, x_3, v_2) \wedge \ldots \wedge p(v_{n-2}, x_n, v_{n-1}) \wedge \\ p(y_1, y_2, u_1) \wedge p(u_1, y_3, u_2) \wedge \ldots \wedge p(u_{n-2}, y_n, u_{n-1}) \wedge \\ E(w, v_{n-1}, u_{n-1}))\end{aligned}$$

For both $N_n(w)$ and $E_n(x_1, \ldots, x_n, y_1, \ldots, y_n)$ we assume that the bound variables are selected so that variable clashes are avoided. For instance, we can assume that the infinite lists $\langle y_i \rangle_{i<\omega}$, $\langle v_i \rangle_{i<\omega}$, etc., are disjoint, and none of them contains z or w. The symbol p will be interpreted as a pairing operator that allows arbitrarily long tuples to be encoded as ordered pairs. The need for N_n will become apparent below, but intuitively it functions as a marker that tells how far to decode a pair into a longer tuple, i.e., whether a certain element c satisfying $p(a, b, c)$ should be taken to represent itself or the pair $\langle a, b \rangle$. Finally, let P be the set containing the sentences of the following forms, where R ranges over relation symbols of ρ, and l is the arity of R:

$$\begin{aligned}\forall x_1 \ldots \forall x_n \forall y_1 \ldots \forall y_n (E_n(x_1, \ldots, x_n, y_1, \ldots, y_n) \supset \\ \bigwedge_{\langle i_1, \ldots, i_l \rangle \in \{1, \ldots, n\}^l} (R(x_{i_1}, \ldots, x_{i_l}) \rightleftharpoons R'(y_{i_1}, \ldots, y_{i_l})))\end{aligned}$$

$$\forall x_1 \ldots \forall x_n \forall y_1 \ldots \forall y_n (E_n(x_1, \ldots, x_n, y_1, \ldots, y_n) \supset \bigwedge\!\!\!\!\bigwedge_{0<i<j\leq n}((x_i = x_j) \rightleftharpoons (y_i = y_j)))$$

$$\exists x_1 \exists x_2 \exists y_1 \exists y_2 \; E_2(x_1, x_2, y_1, y_2)$$

$$\forall x_1 \ldots \forall x_n \forall y_1 \ldots \forall y_n \forall x_{n+1} (E_n(x_1, \ldots, x_n, y_1, \ldots, y_n) \supset \exists y_{n+1} E_{n+1}(x_1, \ldots, x_{n+1}, y_1, \ldots, y_{n+1}))$$

$$\forall x_1 \ldots \forall x_n \forall y_1 \ldots \forall y_n \forall y_{n+1} (E_n(x_1, \ldots, x_n, y_1, \ldots, y_n) \supset \exists x_{n+1} E_{n+1}(x_1, \ldots, x_{n+1}, y_1, \ldots, y_{n+1})).$$

Let P_0 be any finite subset of P. We prove that the countably infinite $\rho \cup \rho'$ model M for $T \cup T' \cup \{\psi, \psi'\}$ can be expanded to a $\rho \cup \rho' \cup \varrho$ model M^0 for P_0. We first define $p^{M^{0+}}$ and $<^{M^{0+}}$. Since $|M|$ is countably infinite, there is an enumeration of $|M|$ without repetitions; $|M| = \{a_n\}_{n<\omega}$. Let $<^{M^{0+}}$ be the (strict) ordering induced by this enumeration. Moreover, let $\langle a_n, a_m, a_k \rangle \in p^{M^{0+}}$ iff $k = 2^n 3^m$. Hence each pair $\langle a_n, a_m \rangle$ is assigned a unique element a_k. We write (a_n, a_m) for this element, and for $n > 2$ we write $(a_{i_1}, \ldots, a_{i_n})$ for $((a_{i_1}, \ldots, a_{i_{n-1}}), a_{i_n})$.

Since $M \models_3 T \cup T'$, we have $M \uparrow \rho \models_3 T$ and $(M \uparrow \rho')^{-'} \models_3 T$. Since T is a maximal consistent set of $\mathcal{L}_\sim$ sentences, by the Ehrenfeucht-Fraïssé result there is some $I = \langle I_n \rangle_{0 \leq n < \omega}$ such that

$$I : M \uparrow \rho \cong_\omega (M \uparrow \rho')^{-'}.$$

Now let m be the largest number such that E_m occurs in P_0. Let a_n be the n'th element of $|M|$, and let b_i and c_i range over arbitrary elements of $|M|$. We let $E^{M^{0+}}$ be the set

$$\{\langle a_n, (b_1, \ldots, b_n), (c_1, \ldots, c_n) \rangle \mid n \leq m, \exists f \in I_{m-n}(f(b_1) = c_1 \wedge \ldots \wedge f(b_n) = c_n)\}.$$

From the definition of E_n we see that

$$M^0 \models_3 E_n(x_1, \ldots, x_n, y_1, \ldots, y_n)[A]$$

iff

$$\langle a_n, (A(x_1), \ldots, A(x_n)), (A(y_1), \ldots, A(y_n)) \rangle \in E^{M^{0+}}.$$

Moreover, since $a_n \neq a_k$ for $n \neq k$, we see that this latter condition holds if *and only if* there is an $f \in I_{m-n}$ such that $f(A(x_i)) = A(y_i)$ for all $i : 1 \leq i \leq n$. But then it immediately follows that M^0 satisfies P_0, since I is an ω-partial isomorphism between $M \uparrow \rho$ and $(M \uparrow \rho')^{-'}$.

We have now shown that every finite subset of $P \cup \{\psi, \sim\psi'\}$ has a model, i.e.,

$$P_0 \cup \{\psi\} \not\models_3 \psi'$$

for all finite $P_0 \subset P$. Hence by countable compactness of $\mathcal{L}^*$ we obtain

$$P \cup \{\psi\} \not\models_3 \psi'.$$

So also the full set $P \cup \{\psi, \sim\psi'\}$ has a model. Since $\mathcal{L}^*$ has the Löwenheim property, $P \cup \{\psi, \sim\psi'\}$ also has a *countable* model N. We show that $(N \uparrow \rho) \cong_p (N \uparrow \rho')^{-'}$. For any n and variable assignment A, let f_{nA} be the (possibly many-valued) function with graph

$$\{\langle A(x_1), A(y_1)\rangle, \ldots, \langle A(x_n), A(y_n)\rangle\}.$$

Then let I be the set

$$\{f_{nA} \mid A,\ n < \omega,\ N \models E_n(x_1, \ldots, x_n, y_1, \ldots, y_n)[A]\}.$$

But now the sentences of P express exactly the condition that the elements of I are proper (not many-valued) functions, and that I is a partial isomorphism. Since $|N|$ is countable, we have shown $(N \uparrow \rho) \cong (N \uparrow \rho')^{-'}$. Since one of these is a model for ψ and the other not, we obtain a contradiction. Hence the proof is complete. □

If we also assume that $\mathcal{L}^*$ is coherent and determinable, then we get strong equivalence in the conclusion:

Lemma. *Let $\mathcal{L}^*$ be a coherent, determinable and countably compact extension of $\mathcal{L}_\sim$ that satisfies the Löwenheim property. Then every $\mathcal{L}^*[\rho]$ sentence is strongly equivalent to an $\mathcal{L}_\sim[\rho]$ sentence.*

Proof: Let $\mathcal{L}^*$ be such an extension, and let $\mathcal{L}^*_\neg$ be a minimal extension of $\mathcal{L}^*$ where for each sentence $\varphi \in \mathcal{L}^*[\rho]$ there is a sentence $\neg\varphi \in \mathcal{L}^*_\neg[\rho]$ such that

$$M \models_{\mathcal{L}^*_\neg} \neg\varphi^{\odot} \quad \text{iff} \quad M \models_{\mathcal{L}^*} \varphi^{\otimes}.$$

Clearly $\mathcal{L}^*_\neg$ has all the properties assumed about $\mathcal{L}^*$. Now let ψ be a sentence of $\mathcal{L}^*[\rho]$. By the previous theorem, there are $\mathcal{L}_\sim[\rho]$ sentences ψ^p and ψ^n that are positively equivalent to ψ and $\neg\psi$, respectively. We recall the definition of the sentence *contr* from chapter 2. Let *compl* be $\neg$*contr*. Clearly $M \models$ *compl* iff M is complete. Since ψ is coherent and determinable, we get

$$\psi^p, \psi^n \models_3 \quad \text{and} \quad compl \models_3 \psi^p, \psi^n,$$

and hence the $\mathcal{L}_\sim[\rho]$ sentence $\sim\sim\psi^p \vee (\sim\psi^p \wedge \sim\psi^n \wedge compl)$ is strongly equivalent to ψ. □

4.3 Relation of $\mathcal{L}$ to $\mathcal{L}_\sim$

In order to prove the combined Lindström and persistence characterization theorem, we have to establish what the relation is between $\mathcal{L}$ and $\mathcal{L}_\sim$. In this section we characterize $\mathcal{L}$ as the persistent fragment of $\mathcal{L}_\sim$. The following auxiliary notions will be useful:

Definition. *An $\mathcal{L}_\sim$ formula φ is* truth-persistent *if for all M, N and A, if $M \ll N$ and $M \models \varphi[A]^+$, then $N \models \varphi[A]^+$. An $\mathcal{L}_\sim$ formula φ is* falsity-persistent *if for all M, N and A, if $M \ll N$ and $M \models \varphi[A]^-$, then $N \models \varphi[A]^-$. Truth-persistence and falsity-persistence relative to generalized structures are defined similarly.*

Hence a formula is persistent iff it is both truth-persistent and falsity-persistent. In characterizing the persistent formulas of $\mathcal{L}_\sim$, it will be helpful to characterize first the truth-persistent and falsity-persistent formulas separately. With little effort, these characterizations can be obtained from results of classical logic and the translation procedures previously considered. But this, also, is how far the translation procedures take us. The function * preserves truth and the dual function $_*$ does, in a sense, preserve falsity, but neither preserves both. The translation procedures can help us to analyze the truth behavior or the falsity behavior of a formula, but not both at the same time. To patch the two separate characterization results together, we shall use the "cut-and-glue" theorem.

We recall the definition of *increasing formulas* from chapter 2.

Lemma. *Let φ be an $\mathcal{L}_\sim$ formula. φ is truth-persistent with respect to generalized structures iff φ^* is increasing.*

Proof: This is so because $M \ll N$ iff $|M^*| = |N^*|$ and $Q^{M^*} \subseteq Q^{N^*}$ for all $Q \in \rho^*$, and because the mapping $M \to M^*$ is *onto.* □

The following theorem from classical logic is now relevant:

Theorem. *An $\mathcal{L}_{\wedge,\forall}[\rho]$ formula is increasing iff it is classically equivalent to an $\mathcal{L}_{\wedge,\forall}[\rho]$ formula where the negation symbol only occurs directly in front of atomic identity formulas.*

Lemma. *An $\mathcal{L}_\sim[\rho]$ formula φ is truth-persistent with respect to generalized structures iff there exists an $\mathcal{L}[\rho]$ formula ψ such that $\models_4 \varphi \equiv \psi$.*

Proof: Equivalence is already established between consecutive items below:

1. φ is truth-persistent with respect to generalized structures.

2. φ^* is increasing.

3. There exists an $\mathcal{L}_{\wedge,\forall}[\rho^*]$ formula χ in which the negation symbol only occurs directly in front of atomic identity formulas, such that $\models_2 \varphi^* \equiv \chi$.
4. There exists an $\mathcal{L}[\rho]$ formula ψ such that $\models_2 \varphi^* \equiv \psi^*$.
5. There exists an $\mathcal{L}[\rho]$ formula ψ such that $\models_4 \varphi \equiv \psi$. □

To establish the corresponding result for proper models, we need the following lemma:

Lemma. *If φ is truth-persistent, then $(\varphi \vee contr)$ is truth-persistent with respect to generalized structures.*

Proof: Suppose φ is truth-persistent, M and N are generalized structures, $M \models (\varphi \vee contr)[A]$ and $M \ll N$. We must show that $N \models (\varphi \vee contr)[A]$.

- If $N \models contr[A]$ then $N \models (\varphi \vee contr)[A]$.
- If $N \not\models contr[A]$ then both M and N are proper. Hence $M \models \varphi[A]$. By truth-persistence of φ with respect to proper models, $N \models \varphi[A]$ and $N \models (\varphi \vee contr)[A]$. □

Theorem. *An $\mathcal{L}_\sim[\rho]$ formula φ is truth-persistent iff there exists an $\mathcal{L}[\rho]$ formula ψ such that $\models_3 \varphi \equiv \psi$.*

Proof: One direction is straightforward. To prove the other direction, suppose φ is truth-persistent. Then $(\varphi \vee contr)$ is truth-persistent with respect to generalized structures. Hence there is an $\mathcal{L}[\rho]$ formula ψ such that $\models_4 ((\varphi \vee contr) \equiv \psi)$. Hence $\models_3 (\varphi \equiv \psi)$. □

Corollary. *An $\mathcal{L}_\sim[\rho]$ formula φ is falsity-persistent iff there exists an $\mathcal{L}[\rho]$ formula χ such that $\models_3 \neg\varphi \equiv \neg\chi$.*

Hence an $\mathcal{L}_\sim[\rho]$ formula φ is persistent iff there exist $\mathcal{L}[\rho]$ formulas ψ and χ such that $\models_3 \varphi \equiv \psi$ and $\models_3 \neg\varphi \equiv \neg\chi$. But we can do better than this: The existence of *two* such $\mathcal{L}$ formulas ψ and χ is equivalent to the existence of a single one filling the role of both. This is an immediate consequence of the "cut-and-glue" theorem. Hence we obtain:

Theorem. *An $\mathcal{L}_\sim[\rho]$ formula is persistent iff it is strongly equivalent to an $\mathcal{L}[\rho]$ formula.*

We also need the corresponding version for sentences, but this follows immediately: If φ is an $\mathcal{L}_\sim$ sentence and ψ is an $\mathcal{L}$ formula, and $\models_3 \varphi \rightleftharpoons \psi$, then also $\models_3 \varphi \rightleftharpoons \chi$, where χ is the existential closure of ψ. Hence we also obtain:

Theorem. *An $\mathcal{L}_\sim[\rho]$ sentence is persistent iff it is strongly equivalent to an $\mathcal{L}[\rho]$ sentence.*

4.4 Maximality of $\mathcal{L}$

We now turn to *persistent* extensions of $\mathcal{L}$. These extensions will lack some of the expressive powers of a non-persistent language like $\mathcal{L}_\sim$, expressive powers used in the proof of the Lindström result above. However, the next lemma shows how to compensate for some of this by passing to *expansions* of a given similarity type.

Lemma. *For every similarity type ρ there is a similarity type $\varrho \supset \rho$, two $\mathcal{L}[\varrho]$ sentences I and J and a translation operator $^{\bullet\bullet}$ from $\mathcal{L}_\sim[\rho]$ to $\mathcal{L}[\varrho]$ such that $I \wedge \sim J \models_3 \varphi^{\bullet\bullet} \equiv \varphi$ for all $\mathcal{L}_\sim[\rho]$ sentences φ, and such that every ρ structure can be expanded to a ϱ structure in which $I \wedge {\sim}J$ is true.*

Proof: Let ρ be any similarity type. We expand ρ to ϱ by adding two new relation symbols $R^{\circ\bullet}$ and $R^{\circ\circ}$ for each relation symbol R of ρ. Let I be the conjunction of $\mathcal{L}[\varrho]$ sentences of the forms

$$\forall x_1 \ldots \forall x_{n_R}(R(x_1, \ldots, x_{n_R}) \vee R^{\circ\bullet}(x_1, \ldots, x_{n_R}))$$

and

$$\forall x_1 \ldots \forall x_{n_R}(\neg R(x_1, \ldots, x_{n_R}) \vee R^{\circ\circ}(x_1, \ldots, x_{n_R})).$$

Similarly, let J be the disjunction of $\mathcal{L}[\varrho]$ sentences of the forms

$$\exists x_1 \ldots \exists x_{n_R}(R(x_1, \ldots, x_{n_R}) \wedge R^{\circ\bullet}(x_1, \ldots, x_{n_R}))$$

and

$$\exists x_1 \ldots \exists x_{n_R}(\neg R(x_1, \ldots, x_{n_R}) \wedge R^{\circ\circ}(x_1, \ldots, x_{n_R})).$$

Note that $I \wedge {\sim}J$ is positively equivalent to the conjunction of $\mathcal{L}_\sim[\varrho]$ sentences of the forms

$$\forall x_1 \ldots \forall x_{n_R}(R^{\circ\bullet}(x_1, \ldots, x_{n_R}) \equiv {\sim}R(x_1, \ldots, x_{n_R}))$$

and

$$\forall x_1 \ldots \forall x_{n_R}(R^{\circ\circ}(x_1, \ldots, x_{n_R}) \equiv {\sim}\neg R(x_1, \ldots, x_{n_R})).$$

If M is a ρ model, let M_ϱ be any ϱ-expansion of M for which

$$R^{\circ\bullet M_\varrho^+} = \widetilde{R^{M^+}} \quad \text{and} \quad R^{\circ\circ M_\varrho^+} = \widetilde{R^{M^-}}.$$

Clearly $M_\varrho \models I \wedge {\sim}J$.

To define the translation function $^{\bullet\bullet}$ we shall need some additional, auxiliary translation functions defined on the full set of $\mathcal{L}_\sim[\rho]$ *formulas*: We define functions $^{\bullet\bullet}$, $^{\bullet\circ}$, $^{\circ\bullet}$ and $^{\circ\circ}$ from $\mathcal{L}_\sim[\rho]$ formulas to $\mathcal{L}[\varrho]$ formulas, satisfying

$(C_{\bullet\bullet})$ $I \wedge {\sim}J \models_3 \varphi^{\bullet\bullet} \equiv \varphi$
$(C_{\bullet\circ})$ $I \wedge {\sim}J \models_3 \varphi^{\bullet\circ} \equiv \neg\varphi$
$(C_{\circ\bullet})$ $I \wedge {\sim}J \models_3 \varphi^{\circ\bullet} \equiv {\sim}\varphi$
$(C_{\circ\circ})$ $I \wedge {\sim}J \models_3 \varphi^{\circ\circ} \equiv {\sim}\neg\varphi$

The definitions are by simultaneous induction on $\mathcal{L}_{\sim}[\rho]$ formulas. The equivalences listed to the right indicate the corresponding steps in an inductive proof for $(C_{\bullet\bullet})$–$(C_{\circ\circ})$.

$(R(x_1,\ldots,x_n))^{\bullet\bullet} = R(x_1,\ldots,x_n)$	
$(R(x_1,\ldots,x_n))^{\bullet\circ} = \neg R(x_1,\ldots,x_n)$	
$(R(x_1,\ldots,x_n))^{\circ\bullet} = R^{\circ\bullet}(x_1,\ldots,x_n)$	
$(R(x_1,\ldots,x_n))^{\circ\circ} = R^{\circ\circ}(x_1,\ldots,x_n)$	
$(x=y)^{\bullet\bullet} = (x=y)^{\circ\circ} = (x=y)$	${\sim}\neg(x=y) \equiv (x=y)$
$(x=y)^{\bullet\circ} = (x=y)^{\circ\bullet} = \neg(x=y)$	${\sim}(x=y) \equiv \neg(x=y)$
$(\exists x\varphi)^{\bullet\bullet} = \exists x\varphi^{\bullet\bullet}$	$\exists x\varphi \equiv \exists x\varphi$
$(\exists x\varphi)^{\bullet\circ} = \forall x\varphi^{\bullet\circ}$	$\neg\exists x\varphi \equiv \forall x\neg\varphi$
$(\exists x\varphi)^{\circ\bullet} = \forall x\varphi^{\circ\bullet}$	${\sim}\exists x\varphi \equiv \forall x{\sim}\varphi$
$(\exists x\varphi)^{\circ\circ} = \exists x\varphi^{\circ\circ}$	${\sim}\neg\exists x\varphi \equiv \exists x{\sim}\neg\varphi$
$(\varphi \vee \psi)^{\bullet\bullet} = (\varphi^{\bullet\bullet} \vee \psi^{\bullet\bullet})$	$(\varphi \vee \psi) \equiv (\varphi \vee \psi)$
$(\varphi \vee \psi)^{\bullet\circ} = (\varphi^{\bullet\circ} \wedge \psi^{\bullet\circ})$	$\neg(\varphi \vee \psi) \equiv (\neg\varphi \wedge \neg\psi)$
$(\varphi \vee \psi)^{\circ\bullet} = (\varphi^{\circ\bullet} \wedge \psi^{\circ\bullet})$	${\sim}(\varphi \vee \psi) \equiv ({\sim}\varphi \wedge {\sim}\psi)$
$(\varphi \vee \psi)^{\circ\circ} = (\varphi^{\circ\circ} \vee \psi^{\circ\circ})$	${\sim}\neg(\varphi \vee \psi) \equiv ({\sim}\neg\varphi \vee {\sim}\neg\psi)$
$(\neg\varphi)^{\bullet\bullet} = \varphi^{\bullet\circ}$	$\neg\varphi \equiv \neg\varphi$
$(\neg\varphi)^{\bullet\circ} = \varphi^{\bullet\bullet}$	$\neg\neg\varphi \equiv \varphi$
$(\neg\varphi)^{\circ\bullet} = (\varphi^{\bullet\bullet} \vee (\varphi^{\circ\bullet} \wedge \varphi^{\circ\circ}))$	${\sim}\neg\varphi \equiv \varphi \vee ({\sim}\varphi \wedge {\sim}\neg\varphi)$
$(\neg\varphi)^{\circ\circ} = (\varphi^{\bullet\circ} \vee (\varphi^{\circ\bullet} \wedge \varphi^{\circ\circ}))$	${\sim}\neg\neg\varphi \equiv \neg\varphi \vee ({\sim}\varphi \wedge {\sim}\neg\varphi)$
$({\sim}\varphi)^{\bullet\bullet} = \varphi^{\circ\bullet}$	${\sim}\varphi \equiv {\sim}\varphi$
$({\sim}\varphi)^{\bullet\circ} = \varphi^{\bullet\bullet}$	$\neg{\sim}\varphi \equiv \varphi$
$({\sim}\varphi)^{\circ\bullet} = \varphi^{\bullet\bullet}$	${\sim}{\sim}\varphi \equiv \varphi$
$({\sim}\varphi)^{\circ\circ} = \varphi^{\circ\bullet}$	${\sim}\neg{\sim}\varphi \equiv {\sim}\varphi$

It is straightforward to check that $(C_{\bullet\bullet})$ - $(C_{\circ\circ})$ hold, that all four operations map *sentences* to *sentences*, and that the image of every $\mathcal{L}_{\sim}[\rho]$ sentence is an $\mathcal{L}[\varrho]$ sentence. □

We use this result in two simple applications.

Lemma. *Let $\mathcal{L}^*$ be an extension of $\mathcal{L}$. If $\mathcal{L}^*$ is countably compact, then so is $\mathcal{L}^* \cup \mathcal{L}_{\sim}$.*

Proof: Let ρ be an arbitrary similarity type, and let ϱ, I, J and $^{\bullet\bullet}$ be as described by the previous lemma. For any set Θ of signed $\mathcal{L}_{\sim}[\rho]$ sentences, let $\Theta^{\bullet\bullet}$ be $\{\varphi^{\bullet\bullet} \mid \varphi^+ \in \Theta\} \cup \{(\neg\varphi)^{\bullet\bullet} \mid \varphi^- \in \Theta\}$.

Now suppose

$$\Gamma^*, \Gamma \models_3 \Delta^*, \Delta$$

where Γ^* and Δ^* are countable sets of signed $\mathcal{L}^*[\rho]$ sentences, and Γ and Δ are countable sets of signed $\mathcal{L}_\sim[\rho]$ sentences. By the properties of I, J and $^{\bullet\bullet}$ it follows that

$$I, \Gamma^*, \Gamma^{\bullet\bullet} \models_3 J, \Delta^*, \Delta^{\bullet\bullet}.$$

Since $\mathcal{L}^*$ is countably compact, also

$$I, \Gamma_0^*, (\Gamma_0)^{\bullet\bullet} \models_3 J, \Delta_0^*, (\Delta_0)^{\bullet\bullet}$$

for finite subsets of the corresponding sets. But then we also have

$$\Gamma_0^*, \Gamma_0 \models_3 \Delta_0^*, \Delta_0,$$

since any ρ structure constituting a counterexample to this last assertion can be expanded to a counterexample to the former. □

Lemma. *Let $\mathcal{L}^*$ be an extension of $\mathcal{L}$. If $\mathcal{L}^*$ satisfies the Löwenheim property, then so does $\mathcal{L}^* \cup \mathcal{L}_\sim$.*

Proof: Let M be a any model for similarity type ρ, and let ϱ, I, J and $^{\bullet\bullet}$ be as described in a lemma above. Let N be an expansion of M satisfying $I \wedge {\sim}J$. Then for every $\mathcal{L}_\sim[\rho]$ sentence φ, $N \models \varphi^{\bullet\bullet +}$ iff $M \models \varphi^+$ and $N \models (\neg\varphi)^{\bullet\bullet +}$ iff $M \models \varphi^-$.

By assumption, $\mathcal{L}^*$ satisfies the Löwenheim property. hence there is a countable ϱ model N_0 such that $N \models \phi$ iff $N_0 \models \phi$ for all signed $\mathcal{L}^*[\varrho]$ sentences ϕ. Then $N_0 \models I \wedge {\sim}J$, and it follows that $M \models \phi$ iff $N_0 \upharpoonright \rho \models \phi$ for all signed $(\mathcal{L}^* \cup \mathcal{L}_\sim)[\rho]$ sentences ϕ. □

Theorem. *Let $\mathcal{L}^*$ be a coherent, determinable, persistent and countably compact extension of $\mathcal{L}$ that satisfies the Löwenheim property. Then every $\mathcal{L}^*[\rho]$ sentence is strongly equivalent to an $\mathcal{L}$ sentence.*

Proof: Assume the antencedent. Then $\mathcal{L}^* \cup \mathcal{L}_\sim$ is a coherent, determinable and countably compact extension of $\mathcal{L}_\sim$ that satisfies the Löwenheim property. Hence every $\mathcal{L}^* \cup \mathcal{L}_\sim[\rho]$ sentence is strongly equivalent to an $\mathcal{L}_\sim[\rho]$ sentence. Furthermore, every $\mathcal{L}^*[\rho]$ sentence is persistent by assumption, and therefore strongly equivalent to a *persistent* $\mathcal{L}_\sim[\rho]$ sentence, and thus again to an $\mathcal{L}[\rho]$ sentence. □

4.5 Truth Functional Extensions of $\mathcal{L}$

In a previous section we found that every persistent sentence of $\mathcal{L}_\sim$ is strongly equivalent to an $\mathcal{L}$ sentence. Using the Lindström theorem, we can strengthen this to a general result about extensions $\mathcal{L}_C$ of $\mathcal{L}$, obtained by adding to the quantifier $\exists$ and connectives $\neg$ and $\vee$ of $\mathcal{L}$ a set C of compositional connectives. Given the relative saturation theorem, it is possible to prove this result in a way quite analogous to the proof of the corresponding result for $\mathbf{l}_C$. But when also the Lindström theorem is given, the present strategy is just as simple.

We recall that in the general setting of not necessarily coherent languages, a *compositional* or *truth functional* n-ary connective c has an associated function f_c from $\{1, 0, \wr, \times\}^n$ into $\{1, 0, \wr, \times\}$ such that for all formulas $\varphi_1, \ldots, \varphi_n$, partial models M and variable assignments A

$$[\![c(\varphi_1, \ldots, \varphi_n)]\!]_{M,A} = f_c([\![\varphi_1]\!]_{M,A}, \ldots, [\![\varphi_n]\!]_{M,A}).$$

We first note that a truth functional completeness result for $\mathcal{L}_{\sim,\star,\diamond}$ can be derived directly from its propositional logic counterpart. The following lemma describes the propositional/predicate logic connection that is needed. As part of the notation below, we assume that when $\top \notin C$ then all occurrences in $\varphi(\begin{smallmatrix}\psi_1\\ S_1\end{smallmatrix} \ldots \begin{smallmatrix}\psi_n\\ S_n\end{smallmatrix})$ of the 0-ary connective $\top$, which is a primitive of $\mathbf{l}$, are replaced by $\exists x(x = x)$:

Lemma. *Let φ be an $\mathbf{l}_C[\rho_0]$ sentence, where $\rho_0 = \{S_1, \ldots, S_n\}$. Moreover, let $\psi_1, \ldots, \psi_n$ be $\mathcal{L}_C[\rho]$ formulas. If φ is valid in propositional logic, then $\varphi(\begin{smallmatrix}\psi_1\\ S_1\end{smallmatrix} \ldots \begin{smallmatrix}\psi_n\\ S_n\end{smallmatrix})$ is valid in predicate logic.*

Proof: There is no combination of truth values to $\psi_1, \ldots, \psi_n$ that cannot be realized by $S_1, \ldots, S_n$. □

Theorem. *Let C be a set of compositional connectives. Then for any $\mathcal{L}_C[\rho]$ formula ψ there is an $\mathcal{L}_{\sim,\star,\diamond}[\rho]$ formula φ such that $\models_3 \varphi \rightleftharpoons \psi$.*

Proof: By induction on $\mathcal{L}_C$ formulas ψ. The basis of the induction is trivial.

Clearly if $\models_3 \psi_0 \rightleftharpoons \varphi_0$ then also $\models_3 \exists x\psi_0 \rightleftharpoons \exists x\varphi_0$.

Now suppose ψ is $c(\psi_1, \ldots, \psi_n)$. By the induction hypothesis there are $\mathcal{L}_{\sim,\star,\diamond}$ formulas $\varphi_1, \ldots, \varphi_n$ such that

$$\models_3 c(\psi_1, \ldots, \psi_n) \rightleftharpoons c(\varphi_1, \ldots, \varphi_n).$$

Let $\rho_0 = \{S_1, \ldots, S_n\}$ be a similarity type for propositional logic. The $\mathbf{l}_C[\rho_0]$ sentence $c(S_1, \ldots, S_n)$ defines a function into $\{0, 1, \wr, \times\}$ from the set of ρ_0 models. Hence by the functional completeness result for $\mathbf{l}_{\sim,\star,\diamond}$ relative to propositional logic, there is a $\mathbf{l}_{\sim,\star,\diamond}[\rho_0]$ sentence ψ_0 such that

$$\models_3 c(S_1, \ldots, S_n) \rightleftharpoons \psi_0.$$

By the lemma above,

$$\models_3 c(\varphi_1, \ldots, \varphi_n) \rightleftharpoons \psi_0(\tfrac{\varphi_1}{S_1} \cdots \tfrac{\varphi_n}{S_n}).$$

Hence

$$\models_3 c(\psi_1, \ldots, \psi_n) \rightleftharpoons \psi_0(\tfrac{\varphi_1}{S_1} \cdots \tfrac{\varphi_n}{S_n}). \qquad \square$$

We shall again make use of the correspondence between generalized and classical structures. Again we extend the translations φ^* and φ_*. We recall the definition of $\top$ and $\bot$ as $\exists x(x = x)$ and $\neg\exists x(x = x)$. The additional clauses are the following:

$$\star^* = \star_* = \bot \qquad\qquad \diamond^* = \diamond_* = \top.$$

Let M be a generalized model. Clearly $M \models \star[A]^+$ iff $M^* \models \star^*[A]^+$, $M \models \star[A]^-$ iff $M^* \models \star_*[A]^+$, $M \models \diamond[A]^+$ iff $M^* \models \diamond^*[A]^+$ and $M \models \diamond[A]^-$ iff $M^* \models \diamond_*[A]^+$. Hence we can construct an inductive proof also for the following:

Lemma. *For any $\mathcal{L}_{\sim,\star,\diamond}$ formula φ and generalized model M, $M \models \varphi[A]^+$ iff $M^* \models \varphi^*[A]^+$, and $M \models \varphi[A]^-$ iff $M^* \models \varphi_*[A]^+$.*

Since the mapping from M to M^* is onto the class of ρ^* models, we derive a corresponding result for consequence relations. For sets Γ of signed $\mathcal{L}_{\sim,\star,\diamond}$ sentences, let Γ^* be the set $\{\varphi^* \mid \varphi^+ \in \Gamma\} \cup \{\varphi_* \mid \varphi^- \in \Gamma\}$.

Lemma. *For any sets Γ and Δ of signed $\mathcal{L}_{\sim,\star,\diamond}$ sentences we have $\Gamma \models_3 \Delta$ iff $\Gamma^* \models_2 \Delta^*, contr^*$.*

Proof: Since a generalized model M is proper iff $M \not\models contr$, we have $\Gamma \models_3 \Delta$ iff $\Gamma \models_4 \Delta, contr$. And by the lemma above we have $\Gamma \models_4 \Delta, contr$ iff $\Gamma^* \models_2 \Delta^*, contr^*$. $\square$

Before we can apply the Lindström theorem, we must show that every language of the form $\mathcal{L}_C$ is countably compact and satisfies the Löwenheim property.

Lemma. *Let C be a set of compositional connectives. Then $\mathcal{L}_C$ is countably compact.*

Proof: It is sufficient to prove this for $\mathcal{L}_{\sim,\star,\diamond}$. So suppose $\Gamma \models_3 \Delta$, where Γ and Δ are countable sets of signed $\mathcal{L}_{\sim,\star,\diamond}[\rho]$ sentences. Then

$$\Gamma^* \models_2 \Delta^*, contr^*,$$

i.e.,

$$\Gamma^* \cup \{\neg\varphi \mid \varphi \in \Delta^*\} \cup \{\neg(contr^*)\}$$

has no model. By the standard compactness result for classical logic, also

$$(\Gamma_0)^* \cup \{\neg\varphi \mid \varphi \in (\Delta_0)^*\} \cup \{\neg(contr^*)\}$$

has no model for some finite subsets Γ_0 and Δ_0 of Γ and Δ. Hence

$$(\Gamma_0)^* \models_2 (\Delta_0)^*, contr^*,$$

and

$$\Gamma_0 \models_3 \Delta_0. \qquad \square$$

Lemma. *Let C be a set of compositional connectives. Then $\mathcal{L}_C$ has the Löwenheim property.*

Proof: Again it is sufficient to prove this for $\mathcal{L}_{\sim,\star,\diamond}$. Let M be an arbitrary model. By the Skolem–Löwenheim theorem for classical logic, there is a countable, complete model N_0 such that for all signed $\mathcal{L}_{\wedge,\forall}[\rho^*]$ sentences ϕ, $N_0 \models \phi$ iff $M^* \models \phi$. Since $M \not\models contr$, we get $M^* \not\models contr^*$ and $N_0 \not\models contr^*$. Hence there is a proper partial model M_0 such that $(M_0)^* = N_0$. Clearly M_0 is countable, and $M_0 \models \varphi^+$ iff $N_0 \models \varphi^*$ iff $M^* \models \varphi^*$ iff $M \models \varphi^+$. Similarly, $M_0 \models \varphi^-$ iff $N_0 \models \varphi_*$ iff $M^* \models \varphi_*$ iff $M \models \varphi^-$. $\square$

Theorem. *Let C be a set of compositional connectives. Then every coherent, determinable and persistent sentence of $\mathcal{L}_C[\rho]$ is strongly equivalent to an $\mathcal{L}[\rho]$ sentence.*

Proof: Let C be a set of compositional connectives, and for all ρ let $\mathcal{L}_C^{cdp}[\rho]$ be the set of $\mathcal{L}_C[\rho]$ sentences that are coherent, determinable and persistent. Moreover, for any $\phi \in \mathcal{L}_C^{cdp}[\rho]$ let $M \models_{\mathcal{L}_C^{cdp}} \phi$ iff $M \models_{\mathcal{L}_C} \phi$. Clearly $\mathcal{L}_C^{cdp}$ is a coherent, determinable, persistent and countably compact extension of $\mathcal{L}$ that satisfies the Löwenheim property. Hence every $\mathcal{L}_C^{cdp}[\rho]$ sentence is strongly equivalent to an $\mathcal{L}[\rho]$ sentence. $\square$

It should be noted that the step from the persistence characterization result for $\mathcal{L}_\sim$ to the full coherence-determinability-persistence characterization result for $\mathcal{L}_C$ in general is considerably cheaper than the application of the Lindström theorem may suggest. This step can also be obtained by the same methods as those used in chapter 1. (The basic persistence characterization result for $\mathcal{L}_\sim$ is deeper than its propositional logic counterpart, however. We shall return to this shortly, when we discuss recursiveness.)

There are alternative routes to these characterization results. In this book it has been the policy to reduce as much as possible to classical logic, and then examine what is left. The questions of truth-persistence and falsity-persistence in $\mathcal{L}_\sim$ were easily reduced to the question of increasing formulas in classical logic. What then remained in order to characterize persistence in $\mathcal{L}_\sim$ was the cut-and-glue/relative saturation theorem.

An alternative strategy to prove these characterization results can be extracted from the dissertation of S. Blamey, cf. Blamey (1980). By four separate but closely related Henkin-style interpolation arguments, it can be shown that for all $\varphi \in \mathcal{L}_{\sim,\star,\diamond}[\rho]$,

(i) If φ is coherent, then $\varphi \models_3 \psi$ and $\neg\varphi \models_3 \neg\psi$ for some $\psi \in \mathcal{L}[\rho]$.

(ii) If φ is determinable, then $\psi \models_3 \varphi$ and $\neg\psi \models_3 \neg\varphi$ for some $\psi \in \mathcal{L}[\rho]$.

(iii) If φ is truth-persistent, then $\varphi \models_3 \psi$ and $\psi \models_3 \varphi$ for some $\psi \in \mathcal{L}[\rho]$.

(iv) If φ is falsity-persistent, then $\neg\varphi \models_3 \neg\psi$ and $\neg\psi \models_3 \neg\varphi$ for some $\psi \in \mathcal{L}[\rho]$.

When a sentence φ has all four properties, a strongly equivalent $\mathcal{L}[\rho]$ sentence can be obtained as a combination in $\vee$ and $\neg$ of the four indicated $\mathcal{L}[\rho]$ sentences.

The present general strategy to prove the persistence characterization result for $\mathcal{L}_\sim$ was introduced in Fenstad et al. (1987). The relative saturation theorem was proved by different methods; no use was made of alternative truth definitions. Instead was introduced a sequent calculus for classical logic, without a full contraction rule. The argument was carried out in a proof theoretic setting, but the main ideas can be recognized in the proofs about $\models_\omega$ given here.

With a result like the coherence-determinability-persistence characterization theorem, which states that *for every* such-and-such sentence, *there exists* another sentence with such-and-such properties in relation to the first, the question arises of whether there is a recursiveness in the correspondence. We prove the following theorem:

Theorem. *There is no recursive function that maps the persistent sentences of $\mathcal{L}_\sim$ onto strongly equivalent $\mathcal{L}$ sentences.*

From this, a corresponding result about coherent, determinable and persistent sentences of $\mathcal{L}_{\sim,\star,\diamond}$ trivially follows. We have already observed a lack of recursiveness in relative saturation; there is no recursive function f from pairs of $\mathcal{L}$ sentences to $\mathcal{L}$ sentences, such that if $\varphi \models_2 \psi$ then $\varphi \models_3 f(\varphi, \psi)$ and $\models_3 \neg\psi \equiv \neg f(\varphi, \psi)$. Hence the above theorem follows by the simple, constructive nature of the next proof.

Proposition. *If the persistence characterization theorem for $\mathcal{L}_\sim$ holds, then the relative saturation theorem holds.*

Proof: Suppose $\varphi \models_2 \psi$ for $\mathcal{L}$ formulas φ and ψ. Since φ and ψ are persistent and coherent, also $\neg\psi \models_3 \sim\varphi$. Let χ be the formula

$$(\psi \wedge \neg\psi) \vee \sim\sim(\varphi \vee \psi).$$

Then

$$\models_3 \chi \equiv (\varphi \vee \psi) \quad \text{and} \quad \models_3 \neg\chi \equiv \neg\psi.$$

Hence χ is persistent, so it is strongly equivalent to an $\mathcal{L}$ formula χ'. Now $\models_3 \neg\chi' \equiv \neg\psi$ and $\models_3 \varphi \supset \chi'$. □

A related question concerns the decidability of persistence. We have the following result, which is due to Johan van Benthem:

Theorem. *The set of persistent $\mathcal{L}_\sim$ sentences is not recursive.*

Proof: We show that for any $\mathcal{L}_\sim$ sentence φ, the sentence $(\varphi \vee \sim compl)$ is persistent iff φ is classically valid. The theorem then follows by the undecidability of classical validity.

So first suppose $\models_2 \varphi$. Then $\models_3 (\varphi \vee \sim compl)$, so trivially $(\varphi \vee \sim compl)$ is both truth-persistent and falsity-persistent.

To prove the other direction, suppose $(\varphi \vee \sim compl)$ is persistent and let M be a complete model. Clearly there exists a non-complete model M_0 such that $M_0 \ll M$. Since $M_0 \models (\varphi \vee \sim compl)$, by persistence $M \models (\varphi \vee \sim compl)$. Since M is complete, this implies $M \models \varphi$. □

4.6 Other Extensions of $\mathcal{L}$

The reader will have noticed that our definition of compactness differs from the standard formulation in classical logic. (See the definition below.) The two versions are equivalent for a large class of languages, in particular those that are closed both under an operator corresponding to $\neg$, and under an operator corresponding to $\sim$. However, for *persistent* languages the two versions are not always equivalent. The motivation for the name 'left-compact' should be obvious; this weakening of the compactness condition corresponds to a restriction from '$\Gamma \models_{\mathcal{L}^*} \Delta$' to '$\Gamma \models_{\mathcal{L}^*}$' in the definition of full countable compactness:

Definition. $\mathcal{L}^*$ *is* countably left-compact, *if for every* ρ *and every countable set* Γ *of signed* $\mathcal{L}^*[\rho]$ *sentences, if every finite subset of* Γ *has a model then* Γ *itself has a model.*

We next identify a coherent, determinable, persistent and countably left-compact extension of $\mathcal{L}$ which satisfies the Löwenheim property, but which still contains sentences that are not strongly, or even positively, equivalent to any $\mathcal{L}$ sentence. The example exploits some of our findings about $\models^*$.

Let $\mathcal{L}^+[\rho]$ be $\mathcal{L}[\rho] \cup \{T(\varphi) \mid \varphi \in \mathcal{L}[\rho]\}$. In other words, the sentences of $\mathcal{L}^+$ are obtained from the sentences of $\mathcal{L}$ by adding a unary sentential operator that does not occur embedded. Moreover, let $M \models T(\varphi)^{\odot}$ iff $M \models^* \varphi^{\odot}$. $\mathcal{L}^+$ should be thought of as the union of the two languages $\langle \mathcal{L}, \models \rangle$ and $\langle \mathcal{L}, \models^* \rangle$. The purpose of the symbol T is to introduce a new disjoint copy of the syntax $\mathcal{L}$, for which the generalized strong Kleene truth definition applies.

We first show that there exist $\mathcal{L}^+$ sentences that are not even positively equivalent to any $\mathcal{L}$ sentence: Suppose otherwise. Then for every $\mathcal{L}$ sentence φ there exists an $\mathcal{L}$ sentence ψ such that $\models_3 T(\varphi) \equiv \psi$. From this we deduce $\models_2 \varphi \equiv \psi$, $\models_3 T(\psi) \equiv \psi$, and hence the equivalence $M \models^* \psi \leftrightarrow M \models \psi$. So every $\mathcal{L}$ sentence φ is classically equivalent to a $\models^*$ predictive $\mathcal{L}$ sentence ψ. But this contradicts our findings in the previous chapter.

So $\mathcal{L}^+$ is a strict extension of $\mathcal{L}$. On the other hand, $\mathcal{L}^+$ is clearly coherent, determinable and persistent. Since both $\models$ and $\models^*$ satisfy the Skolem property for truth definitions for $\mathcal{L}$, clearly both $\langle \mathcal{L}, \models \rangle$ and $\langle \mathcal{L}, \models^* \rangle$ satisfy the Skolem property for languages.

Now the union $\mathcal{L}_1 \cup \mathcal{L}_2$ of two languages with the Skolem property will itself have the Skolem property. For any M and any $i \in \{1, 2\}$, let ξ_i be a *cna* such that all substructures of M closed under ξ_i are $\mathcal{L}_i$ equivalent to M. Let ξ be the pointwise union of ξ_1 and ξ_2. ξ is a *cna*, and every substructure of M closed under ξ is closed under both ξ_1 and ξ_2, and is therefore both $\mathcal{L}_1$ equivalent and $\mathcal{L}_2$ equivalent to M, i.e., $\mathcal{L}_1 \cup \mathcal{L}_2$ equivalent to M. Hence, since both $\langle \mathcal{L}, \models \rangle$ and $\langle \mathcal{L}, \models^* \rangle$ satisfy the Skolem property, also their union $\mathcal{L}^+ = \langle \mathcal{L}, \models \rangle \cup \langle \mathcal{L}, \models^* \rangle$ satisfies the Skolem property, and hence also the Löwenheim property.

To prove countable left-compactness, let Γ and Δ be countable sets of signed $\mathcal{L}[\rho]$ sentences (this condition of countability is in fact vacuous for $\mathcal{L}$ and $\mathcal{L}^+$ since only a countable number of such sentences exist), and suppose that $\Gamma \cup \{T(\varphi)^{\odot} \mid \varphi^{\odot} \in \Delta\}$ has no model. Then neither does $\Gamma \cup \Delta$ have a model, since any model of $\varphi^{\odot}$ is a model of $T(\varphi)^{\odot}$. By countable compactness of $\mathcal{L}$ there are finite subsets Γ_0 and Δ_0 of Γ and Δ respectively, such that $\Gamma_0 \cup \Delta_0$ has no model. But then neither can $\Gamma_0 \cup \{T(\varphi)^{\odot} \mid \varphi^{\odot} \in \Delta_0\}$

have a model, since any completion of a counterexample would be a model of $\Gamma_0 \cup \Delta_0$.

To sum up; the language $\mathcal{L}^+ = \langle \mathcal{L}, \models \rangle \cup \langle \mathcal{L}, \models\!\!=^* \rangle$ would constitute a counterexample to the Lindström theorem if 'countably left-compact' were to be substituted for 'countably compact' in the formulation of that theorem. So the Lindström theorem therefore tells us that $\mathcal{L}^+$ is not countably compact. Analogously to $\mathcal{L}^+$, we can also define $\mathcal{L}^\omega = \langle \mathcal{L}, \models \rangle \cup \langle \mathcal{L}, \models\!\!=^*_\omega \rangle$ and $\mathcal{L}^\Box = \langle \mathcal{L}, \models \rangle \cup \langle \mathcal{L}, \models_\Box \rangle$. In a similar way, we can deduce that $\mathcal{L}^\omega$ is not countably compact: $\mathcal{L}^\omega$ is coherent, determinable and persistent; since both $\langle \mathcal{L}, \models \rangle$ and $\langle \mathcal{L}, \models\!\!=^*_\omega \rangle$ satisfy the Skolem property, so does $\mathcal{L}^\omega$. Hence $\mathcal{L}^\omega$ also satisfies the Löwenheim property. Since not every $\mathcal{L}$ sentence is classically equivalent to a $\models\!\!=^*_\omega$ predictive sentence, it follows that not every $\mathcal{L}^\omega$ sentence is positively equivalent to an $\mathcal{L}$ sentence. Hence neither $\mathcal{L}^\omega$ can be countably compact. For $\mathcal{L}^\Box$, things are different, since $\langle \mathcal{L}, \models_\Box \rangle$ and hence $\mathcal{L}^\Box$ does not satisfy the Löwenheim property. So here we cannot conclude directly from the Lindström theorem that $\mathcal{L}^\Box$ is not countably compact.

However, it is not difficult to see that countable compactness fails for $\mathcal{L}^\Box$ as well. $\mathcal{L}^\Box$ can be viewed as a minimal extension of $\mathcal{L}$ for which $\Box\varphi \in \mathcal{L}^\Box[\rho]$ if $\varphi \in \mathcal{L}[\rho]$, and for which $M \models \Box\varphi^\odot$ iff $M \models_\Box \varphi^\odot$. We recall our findings about the sentences $\{\phi_n\}_{n<\omega}$ in the proof that a *uniform* saturation theorem fails. We have

$$\Box\phi_3 \models_3 \{\phi_n\}_{n<\omega}$$

but not

$$\Box\phi_3 \models_3 \{\phi_n\}_{n<N}$$

for any $N < \omega$. Hence $\mathcal{L}^\Box$ is not countably compact.

Implications from the Lindström theorem are not so easily derived about the truth definitions $\models\!\!=^*$, $\models\!\!=^*_\omega$ and $\models_\Box$ themselves, since none of the languages $\langle \mathcal{L}, \models\!\!=^* \rangle$, $\langle \mathcal{L}, \models\!\!=^*_\omega \rangle$ and $\langle \mathcal{L}, \models_\Box \rangle$ are extensions of the language $\mathcal{L}$. The proof of the Lindström theorem makes essential use of the assumption that every sentence of $\mathcal{L}^*$ is compact together with the full set of $\mathcal{L}$ sentences.

A further extension $\mathcal{L}_\Box$ of $\mathcal{L}^\Box$ is obtained if we allow arbitrary embeddings of an operator $\Box$ satisfying the equivalence

$$M \models \Box\varphi^\odot \quad \text{iff} \quad (N \models \varphi^\odot \quad \text{for all completions } N \text{ of } M).$$

Nestings of $\Box$ do not represent a particularly exciting innovation, since for instance

$$\models_3 \Box\Box\varphi \rightleftharpoons \Box\varphi,$$

and in general

$$\models_3 \Box\varphi \rightleftharpoons \Box\varphi',$$

where φ' is φ with all occurrences of $\Box$ deleted. Embeddings of the sort $\exists x \Box \varphi$ are more interesting. Note that

$$\Box \exists x \varphi \not\models_3 \exists x \Box \varphi.$$

We show that the addition of $\Box$ to $\mathcal{L}$ is equivalent to the addition of a symbol $\frown$ for "forcing negation:"

$$M \models \frown\varphi^{\odot} \quad \text{iff} \quad (N \models \varphi^{\odot} \quad \text{for no } N \text{ such that } M \ll N).$$

This is the strongest possible clause for negation that still ensures reliability. Let $\mathcal{L}_{\frown}$ be $\mathcal{L}$ with forcing negation added. Clearly $\mathcal{L}_{\frown}$ is coherent, determinable and persistent. Hence also the following equivalence holds:

$$M \models \frown\varphi^{\odot} \quad \text{iff} \quad (N \models \varphi^{\otimes} \quad \text{for all completions } N \text{ of } M).$$

It follows that we can define $\frown\varphi$ as $\Box\neg\varphi$ and $\Box\varphi$ as $\frown\neg\varphi$, and that $\mathcal{L}_{\Box}$ and $\mathcal{L}_{\frown}$ have the same expressive power.

This seems to be as close as we can get to forcing in the present framework. But the differences are many, and important. First, a forcing model has *more structure* in that it contains a restricted set of entities, each corresponding to one of our partial models, and it identifies a primitive accessibility relation on this set, roughly corresponding to our relation $\ll$ of informational extension. To fit this into our framework, we substituted $\ll$ for the primitive accessibility relation, and let the restricted set/class of entities be the whole class of models in our sense.

And equally important, the accessibility relation of forcing models allows for the extension of the domain of individuals. By substituting $\ll$ for the accessibility relation, this is no longer allowed. This is a serious restriction; many interesting phenomena now disappear.

There are other differences as well; notably in the forcing models partiality is introduced in an asymmetrical way. For the relation symbols only *positive* extensions are defined. These may grow with added information. There are no analogous *negative extensions.* We have considered partial models with both positive and negative extensions. This has motivated the alternative negation symbol $\neg$ with a "flip-flop" interpretation.

To sum up; there are certain similarities between the language $\mathcal{L}_{\Box}$ for our models and the forcing interpretation of $\mathcal{L}$, as used to describe intuitionistic forcing models. Hence it could happen that further studies of $\mathcal{L}_{\Box}$ could benefit from the accumulated knowledge about forcing; insights could even pass in the opposite direction. But differences are many, and substantial. Note in particular that the move from $\ll$ to a primitive relation on a restricted set is somewhat reminiscent of the move from proper second order models to *generalized* such models, where the universal second order quantifiers only range over a designated subset of the full powerset of the ground domain.

Further Directions

This book does not tell the whole story about partial model theory, even less the whole story about partial *logic*. To begin with, the present concept of *partiality* is not very general. For instance, it does not give us anything like a partial identity theory. Future studies should try to push the concept of partiality further, and ask similar questions about more radically partial model theories.

Thus the generalization process from completeness to partiality can be carried further, but more generality is also afforded if we start out with classical models of a more general format. Similarity types for first order logic usually contain also function symbls, with individual constants as special cases. These can be expanded further with higher order relation symbols, and, in the case of type theory, with symbols for the objects of any type. When the questions of this book are raised again for partial type theory, many new distinctions and interesting complications could very well arise.

But many questions also remain about our own modest framework, with our own modest form of partiality. We have studied some specific truth definitions with certain appealing properties, but little has been said about which alternatives may exist. It would be interesting to see some more systematic results about truth definitions in general. For instance, can we identify some reasonable conditions that distinguish the quadruple $\models$, $\models_w^2$, $\models^*$, $\models_\Box$ from all the alternatives? And what exactly is the relation between teh Skolem property and the Löwenheim property for truth definitions? Are they equivalent? If not, is there an important class of truth definitions for which they coincide? We have characterized $\models^*$ as the strongest reliable truth definition with the Skolem property; is it also the strongest reliable truth definition with the Löwenheim property? We have given a sufficient algebraic condition for semanitcal equivalence relative to $\models^*$; can we find a characterization?

Consequence relations have not yet been a topic in this book. Relations resembling $\models_3$ and $\models_4$ have already been axiomatized by a number of authors, cf. for instance Blamey (1980, 1986), Kamp (1983), Langholm (1984), Fenstad et al. (1987). For an arbitrary truth definition $\models_x$, we define $\Gamma \models_x \Delta$ to hold *iff* for every model M and variable assignment A such that $M \models_x \varphi[A]^\odot$ for all $\varphi^\odot \in \Gamma$, $M \models_< \psi[A]^\odot$ for some $\psi^\odot \in \Delta$. When Δ is a singleton, the consequence relations $\models_\Box$, $\models^*$ and $\models^*_w$ all coincide with the classical conequence relation $\models_2$.

Neither of $\Gamma \models_\Box \Delta$, $\Gamma \models^* \Delta$ or $\Gamma \models^*_\omega \Delta$ coincides with $\Gamma \models_2 \Delta$ in general, however. This is rather trivial, for clearly $\models_2 \exists x P(x), \exists x \neg P(x)$, while $\not\models_\Box \exists x P(x), \exists x \neg P(x)$. Similarly for $\models^*$ and $\models^*_\omega$. This corresponds to the familiar modal logic non-equivalence between $\Box(\varphi \vee \psi)$ and $\Box\varphi \vee \Box\psi$.

We have seen that $\langle \mathcal{L}, \models \rangle \cup \langle \mathcal{L}, \models^* \rangle$ and hence $\langle \mathcal{L}, \models^* \rangle$ is countably left-compact. By similar arguments, so are $\langle \mathcal{L}, \models_\Box \rangle$ and $\langle \mathcal{L}, \models^*_\omega \rangle$. Whether or not they are countably compact, we shall leave for future research to decide.

We note that the consequence relations $\models_\Box$ and $\models^*$ will have to be distinct, since a general implication from $\Gamma \models^* \Delta$ to $\Gamma \models_\Box \Delta$ would imply that $\models_\Box$ has the Löwenheim property: For any M we have $\Gamma_M \not\models_\Box \Gamma_{\sim M}$, where $\Gamma_M = \{\phi \in \mathcal{L}[\rho] \mid M \models_\Box \phi\}$ and $\Gamma_{\sim M} = \{\phi \in \mathcal{L}[\rho] \mid M \not\models_\Box \phi\}$. Hence *if* such an implication were to hold, then we would also get $\Gamma_M \not\models^* \Gamma_{\sim M}$, and there would be a countable counterexample since $\models^*$ satisfies the Löwenheim property. Since the truth definitions $\models_\Box$ and $\models^*$ coincide on such structures, this countable structure is also a counterexample to $\Gamma_M \models_\Box \Gamma_{\sim M}$, and hence $\langle \mathcal{L}, \models_\Box \rangle$ equivalent to M.

On the other hand, $\Gamma \models_\Box \Delta$ implies $\Gamma \models^* \Delta$. For suppose $\Gamma \not\models^* \Delta$. Then there is a countable counterexample. On such structures, the truth definitions $\models_\Box$ and $\models^*$ coincide. Hence this countable structure is also a counterexample to $\Gamma \models_\Box \Delta$.

Also in chapter 4 there are many open questions. The Lindström theorem is an important result, but it should not keep us from looking also at stronger languages and their relation to coherence, determinability and persistance. In particular, applications in natural language semantics has motivated a recent interest in generalized quantifiers and various classes of such quantifiers, cf. for instance Gärdenfors (1987). Many characterization results have been proved, and it is of interest to see how these results extend to the partial model theory of this book.

References

Barwise, J. 1974. Axioms for Abstract Model Theory. *Annals of Mathematical Logic*, 7:221–265.

Barwise, J. 1977. An Introduction to First-Order Logic. In J. Barwise, (Ed.), *Handbook of Mathematical Logic.* Amsterdam: North-Holland.

Barwise, J. 1981. Scenes and Other Situations. *Journal of Philosophy*, 78:369–397.

Barwise, J. 1985. The Situation in Logic II: Conditionals and Conditional Information. Report No. CSLI–85–21. Stanford: CSLI.

Barwise, J. and J. Etchemendy. 1987. *The Liar: An Essay on Truth and Circular Propositions.* Oxford: Oxford University Press.

Barwise, J. and J. Perry. 1983. *Situations and Attitudes.* Cambridge, Mass.: Bradford Books/MIT Press.

Blamey, S. 1915. Über Möglichkeiten im Relativkalkül. *Mathematische Annalen*, 76:447–470.

Blamey, S. 1980. *Partial-Valued Logic.* Doctoral dissertation, Oxford University.

Blamey, S. 1986. Partial Logic. In D. Gabbay and F. Guenthner, (Eds.), *Handbook of Philosophical Logic.* Vol. III, *Alternatives in Classical Logic.* Dordrecht: D. Reidel.

Bochvar, D. A. 1981. On a Three-valued Logical Calculus and its Application to the Analysis of the Paradoxes of the Classical Extended Functional Calculus. *History and Philosophy of Logic*, 2:87–112. First published in Russian, *Mathematischeskii sbornik*, 1937, 4(46):247–308.

Chang, C. C. and H. J. Keisler. 1973. *Model Theory.* Amsterdam: North Holland.

Dunn, M. 1986. Relevance Logic and Entailment. In D. Gabbay and F. Guenthner, eds., *Handbook of Philosophical Logic, Vol III: Alternatives in Classical Logic.* Dordrecht: D. Reidel.

Ebbinghaus, H.-D. 1985. Extended Logics: The General Framework. In J. Barwise and S. Feferman, (Eds.), *Model-Theoretic Logics.* Berlin: Springer-Verlag.

Ehrenfeucht, A. 1957. Applications of games to some problems of mathematical logic. *Bull. Acad. Polon. Sci. Sér. Sci. Math. Astronom. Phys.,* 5: 35–37.

Feferman, S. 1960. Some Recent Work of Ehrenfeucht and Fraïssé. *Summaries of Talks Presented at the Summer Institute for Symbolic Logic, Cornell University 1957*, 2d ed., Institute for Defense Analyses, 1960, 201–209.

Feferman, S. 1968. Lectures on Proof Theory. Springer Lecture Notes in Mathematics No. 70. Berlin:Springer-Verlag, 1–107.

Feferman, S. 1984. Toward Useful Type-Free Theories. I. *Journal of Symbolic Logic*, 49:75–111.

Fenstad, J. E. et al. 1987. *Situations, Language and Logic.* Dordrecht: D. Reidel.

Fitting, M. C. 1969. *Intuitionistic Logic, Model Theory and Forcing.* Amsterdam: North-Holland.

Flum, J. 1985. Characterizing Logics. In J. Barwise and S. Feferman, (Eds.), *Model-Theoretic Logics.* Berlin: Springer-Verlag.

van Fraassen, B. 1968. Presupposition, Implication and Self-Reference. *Journal of Philosophy*, 65:135–152.

Fraïssé, R. J. 1955. Sur quelques classifications des relations, basées sur des isomorphismes restreints. I. Étude générale. II. Application aux relations d'ordres. *Alger-Mathématiques*, 2:16–60, 273–295.

Gärdenfors, P., (Ed.), 1987. *Generalized Quantifiers. Linguistic and Logical Approaches.* Dordrecht: D. Reidel.

Gilmore, P. C. 1974. The Consistency of Partial Set Theory without Extensionality. In *Axiomatic Set Theory,* Proceedings of Symposia in Pure Mathematics, Vol. 13, Part II. Providence, R.I.: American Mathematical Society.

Kamp, H. 1983. A Scenic Tour through the Land of Naked Infinitives. To appear in *Linguistics and Philosophy.*

Kleene, S. 1938. On Notation for Ordinal Numbers. *Journal of Symbolic Logic*, 3:150–155.

Kleene, S. 1952. *Introduction to Metamathematics.* Princeton, N.J.: Van Nostrand.

Langholm, T. 1984. Some Tentative Systems Relating to Situation Semantics. In *Report on an Oslo Seminar in Logic and Linguistics,* Preprint Series 9, Mathematical Institute, University of Oslo.

Levy, A. 1979. *Basic Set Theory.* Berlin: Springer-Verlag.

Lindström, P. 1966. On Characterizability in $\mathcal{L}_{\omega_1\omega_0}$. *Theoria*, 32:165–171.

Lindström, P. 1969. On Extensions of Elementary Logic. *Theoria*, 35:1–11.

Löwenheim, L. 1915. Über Möglichkeiten in Relativkalkül. *Mathematische Annalen*, 76:447–470.

Lyndon, R. 1959a. An Interpolation Theorem in the Predicate Calculus. *Pacific Journal of Mathematics*, 9:129–142.

Lyndon, R. 1959b. Properties Preserved under Homomorphism. *Pacific Journal of Mathematics*, 9:143–154.

Skolem, T. 1920. Logisch-kombinatorische Untersuchungen über die Erfüllbarkeit und Beweisbarkeit mathematischen Sätze nebst einem Theoreme über dichte Mengen. *Skriffer, Videnskabsakademiet i Kristiana*, I, No. 4, 1–36. Reprinted in J. E. Fenstad, (Ed.), 1970, *Th. Skolem, Selected Works in Logic*, Oslo: Universitetsforlaget.

Urquhart, A. 1986. Many-Valued Logic. In D. Gabbay and F. Guenthner, (Eds.), *Handbook of Philosophical Logic.* Vol. III, *Alternatives in Classial Logic.* Dordrecht: D. Reidel.

Van Benthem, J. 1984. *Partiality and Nonmonotonicity in Classical Logic.* Report No. CSLI–84–12. Stanford: CSLI.

Van Benthem, J. 1985. *A Manual of Intensional Logic.* CSLI Lecture Notes No. 1. Stanford: CSLI.

Visser, A. 1984. Four Valued Semantics and the Liar. *Journal of Philosophical Logic*, 13:181–212.

Woodruff, P. 1984. On Supervaluations in Free Logic. *Journal of Symbolic Logic*, 49:943–950.

Index of Symbols

General Index

CSLI Publications

Reports

The following titles have been published in the CSLI Reports series. These reports may be obtained from CSLI Publications, Ventura Hall, Stanford University, Stanford, CA 94305-4115.

The Situation in Logic–I Jon Barwise CSLI–84–2 ($*2.00*)

Coordination and How to Distinguish Categories Ivan Sag, Gerald Gazdar, Thomas Wasow, and Steven Weisler CSLI–84–3 ($*3.50*)

Belief and Incompleteness Kurt Konolige CSLI–84–4 ($*4.50*)

Equality, Types, Modules and Generics for Logic Programming Joseph Goguen and José Meseguer CSLI–84–5 ($*2.50*)

Lessons from Bolzano Johan van Benthem CSLI–84–6 ($*1.50*)

Self-propagating Search: A Unified Theory of Memory Pentti Kanerva CSLI–84–7 ($*9.00*)

Reflection and Semantics in LISP Brian Cantwell Smith CSLI–84–8 ($*2.50*)

The Implementation of Procedurally Reflective Languages Jim des Rivières and Brian Cantwell Smith CSLI–84–9 ($*3.00*)

Parameterized Programming Joseph Goguen CSLI–84–10 ($*3.50*)

Morphological Constraints on Scandinavian Tone Accent Meg Withgott and Per-Kristian Halvorsen CSLI–84–11 ($*2.50*)

Partiality and Nonmonotonicity in Classical Logic Johan van Benthem CSLI–84–12 ($*2.00*)

Shifting Situations and Shaken Attitudes Jon Barwise and John Perry CSLI–84–13 ($*4.50*)

Aspectual Classes in Situation Semantics Robin Cooper CSLI–85–14-C ($*4.00*)

Completeness of Many-Sorted Equational Logic Joseph Goguen and José Meseguer CSLI–84–15 ($*2.50*)

Moving the Semantic Fulcrum Terry Winograd CSLI–84–17 ($*1.50*)

On the Mathematical Properties of Linguistic Theories C. Raymond Perrault CSLI–84–18 ($*3.00*)

A Simple and Efficient Implementation of Higher-order Functions in LISP Michael P. Georgeff and Stephen F.Bodnar CSLI–84–19 ($*4.50*)

On the Axiomatization of "if-then-else" Irène Guessarian and José Meseguer CSLI–85–20 ($*3.00*)

The Situation in Logic–II: Conditionals and Conditional Information Jon Barwise CSLI–84–21 ($*3.00*)

Principles of OBJ2 Kokichi Futatsugi, Joseph A. Goguen, Jean-Pierre Jouannaud, and José Meseguer CSLI–85–22 ($*2.00*)

Querying Logical Databases Moshe Vardi CSLI–85–23 ($*1.50*)

Computationally Relevant Properties of Natural Languages and Their Grammar Gerald Gazdar and Geoff Pullum CSLI–85–24 ($*3.50*)

An Internal Semantics for Modal Logic: Preliminary Report Ronald Fagin and Moshe Vardi CSLI–85–25 ($*2.00*)

The Situation in Logic–III: Situations, Sets and the Axiom of Foundation Jon Barwise CSLI–85–26 ($*2.50*)

Semantic Automata Johan van Benthem CSLI–85–27 ($*2.50*)

Restrictive and Non-Restrictive Modification Peter Sells CSLI–85–28 ($*3.00*)

Institutions: Abstract Model Theory for Computer Science J. A. Goguen and R. M. Burstall CSLI–85–30 ($*4.50*)

A Formal Theory of Knowledge and Action Robert C. Moore CSLI–85–31 ($*5.50*)

Finite State Morphology: A Review of Koskenniemi (1983) Gerald Gazdar CSLI–85–32 ($*1.50*)

The Role of Logic in Artificial Intelligence Robert C. Moore CSLI–85–33 ($*2.00*)

Applicability of Indexed Grammars to Natural Languages Gerald Gazdar CSLI–85–34 ($*2.00*)

Commonsense Summer: Final Report Jerry R. Hobbs, et al CSLI–85–35 ($*12.00*)

Limits of Correctness in Computers Brian Cantwell Smith CSLI–85–36 ($*2.50*)

On the Coherence and Structure of Discourse Jerry R. Hobbs CSLI–85–37 ($*3.00*)

The Coherence of Incoherent Discourse Jerry R. Hobbs and Michael H. Agar CSLI–85–38 ($*2.50*)

The Structures of Discourse Structure Barbara Grosz and Candace L. Sidner CSLI–85–39 ($*4.50*)

A Complete, Type-free "Second-order" Logic and Its Philosophical Foundations Christopher Menzel CSLI–86–40 ($*4.50*)

Possible-world Semantics for Autoepistemic Logic Robert C. Moore CSLI–85–41 ($*2.00*)

Deduction with Many-Sorted Rewrite José Meseguer and Joseph A. Goguen CSLI–85–42 ($*1.50*)

On Some Formal Properties of Metarules Hans Uszkoreit and Stanley Peters CSLI–85–43 ($*1.50*)

Language, Mind, and Information John Perry CSLI–85–44 ($*2.00*)

Constraints on Order Hans Uszkoreit CSLI–86–46 ($*3.00*)

Linear Precedence in Discontinuous Constituents: Complex Fronting in German Hans Uszkoreit CSLI–86–47 ($*2.50*)

A Compilation of Papers on Unification-Based Grammar Formalisms, Parts I and II Stuart M. Shieber, Fernando C.N. Pereira, Lauri Karttunen, and Martin Kay CSLI–86–48 ($*10.00*)

An Algorithm for Generating Quantifier Scopings Jerry R. Hobbs and Stuart M. Shieber CSLI–86–49 ($*2.50*)

Verbs of Change, Causation, and Time Dorit Abusch CSLI–86–50 ($*2.00*)

Noun-Phrase Interpretation Mats Rooth CSLI–86–51 ($*2.00*)

Noun Phrases, Generalized Quantifiers and Anaphora Jon Barwise CSLI–86–52 ($*2.50*)

Circumstantial Attitudes and Benevolent Cognition John Perry CSLI–86–53 ($*1.50*)

A Study in the Foundations of Programming Methodology: Specifications, Institutions, Charters and Parchments Joseph A. Goguen and R. M. Burstall CSLI–86–54 ($*2.50*)

Quantifiers in Formal and Natural Languages Dag Westerståhl CSLI–86–55 ($*7.50*)

Intentionality, Information, and Matter Ivan Blair CSLI–86–56 ($*3.00*)

Graphs and Grammars William Marsh CSLI–86–57 ($*2.00*)

Computer Aids for Comparative Dictionaries Mark Johnson CSLI–86–58 ($*2.00*)

The Relevance of Computational Linguistics Lauri Karttunen CSLI–86–59 ($*2.50*)

Grammatical Hierarchy and Linear Precedence Ivan A. Sag CSLI–86–60 ($*3.50*)

D-PATR: A Development Environment for Unification-Based Grammars Lauri Karttunen CSLI–86–61 (*$4.00*)

A Sheaf-Theoretic Model of Concurrency Luís F. Monteiro and Fernando C. N. Pereira CSLI–86–62 (*$3.00*)

Discourse, Anaphora and Parsing Mark Johnson and Ewan Klein CSLI–86–63 (*$2.00*)

Tarski on Truth and Logical Consequence John Etchemendy CSLI–86–64 (*$3.50*)

The LFG Treatment of Discontinuity and the Double Infinitive Construction in Dutch Mark Johnson CSLI–86–65 (*$2.50*)

Categorial Unification Grammars Hans Uszkoreit CSLI–86–66 (*$2.50*)

Generalized Quantifiers and Plurals Godehard Link CSLI–86–67 (*$2.00*)

Radical Lexicalism Lauri Karttunen CSLI–86–68 (*$2.50*)

Understanding Computers and Cognition: Four Reviews and a Response Mark Stefik, Editor CSLI–87–70 (*$3.50*)

The Correspondence Continuum Brian Cantwell Smith CSLI–87–71 (*$4.00*)

The Role of Propositional Objects of Belief in Action David J. Israel CSLI–87–72 (*$2.50*)

From Worlds to Situations John Perry CSLI–87–73 (*$2.00*)

Two Replies Jon Barwise CSLI–87–74 (*$3.00*)

Semantics of Clocks Brian Cantwell Smith CSLI–87–75 (*$2.50*)

Varieties of Self-Reference Brian Cantwell Smith CSLI–87–76 (*Forthcoming*)

The Parts of Perception Alexander Pentland CSLI–87–77 (*$4.00*)

Topic, Pronoun, and Agreement in Chicheŵa Joan Bresnan and S. A. Mchombo CSLI–87–78 (*$5.00*)

HPSG: An Informal Synopsis Carl Pollard and Ivan A. Sag CSLI–87–79 (*$4.50*)

The Situated Processing of Situated Language Susan Stucky CSLI–87–80 (*Forthcoming*)

Muir: A Tool for Language Design Terry Winograd CSLI–87–81 (*$2.50*)

Final Algebras, Cosemicomputable Algebras, and Degrees of Unsolvability Lawrence S. Moss, José Meseguer, and Joseph A. Goguen CSLI–87–82 (*$3.00*)

The Synthesis of Digital Machines with Provable Epistemic Properties Stanley J. Rosenschein and Leslie Pack Kaelbling CSLI–87–83 (*$3.50*)

Formal Theories of Knowledge in AI and Robotics Stanley J. Rosenschein CSLI–87–84 (*$1.50*)

An Architecture for Intelligent Reactive Systems Leslie Pack Kaelbling CSLI–87–85 (*$2.00*)

Order-Sorted Unification José Meseguer, Joseph A. Goguen, and Gert Smolka CSLI–87–86 (*$2.50*)

Modular Algebraic Specification of Some Basic Geometrical Constructions Joseph A. Goguen CSLI–87–87 (*$2.50*)

Persistence, Intention and Commitment Phil Cohen and Hector Levesque CSLI–87–88 (*$3.50*)

Rational Interaction as the Basis for Communication Phil Cohen and Hector Levesque CSLI–87–89 ()

An Application of Default Logic to Speech Act Theory C. Raymond Perrault CSLI–87–90 (*$2.50*)

Models and Equality for Logical Programming Joseph A. Goguen and José Meseguer CSLI–87–91 (*$3.00*)

Order-Sorted Algebra Solves the Constructor-Selector, Mulitple Representation and Coercion Problems Joseph A. Goguen and José Meseguer CSLI–87–92 (*$2.00*)

Extensions and Foundations for Object-Oriented Programming Joseph A. Goguen and José Meseguer CSLI–87–93 ($*3.50*)

L3 Reference Manual: Version 2.19 William Poser CSLI–87–94 ($*2.50*)

Change, Process and Events Carol E. Cleland CSLI–87–95 (*Forthcoming*)

One, None, a Hundred Thousand Specification Languages Joseph A. Goguen CSLI–87–96 ($*2.00*)

Constituent Coordination in HPSG Derek Proudian and David Goddeau CSLI–87–97 ($*1.50*)

A Language/Action Perspective on the Design of Cooperative Work Terry Winograd CSLI–87–98 ($*2.50*)

Implicature and Definite Reference Jerry R. Hobbs CSLI–87–99 ($*1.50*)

Thinking Machines: Can There be? Are we? Terry Winograd CSLI–87–100 ($*2.50*)

Situation Semantics and Semantic Interpretation in Constraint-based Grammars Per-Kristian Halvorsen CSLI–87–101 ($*1.50*)

Category Structures Gerald Gazdar, Geoffrey K. Pullum, Robert Carpenter, Ewan Klein, Thomas E. Hukari, Robert D. Levine CSLI–87–102 ($*3.00*)

Cognitive Theories of Emotion Ronald Alan Nash CSLI–87–103 ($*2.50*)

Toward an Architecture for Resource-bounded Agents Martha E. Pollack, David J. Israel, and Michael E. Bratman CSLI–87–104 ($*2.00*)

On the Relation Between Default and Autoepistemic Logic Kurt Konolige CSLI–87–105 ($*3.00*)

Three Responses to Situation Theory Terry Winograd CSLI–87–106 ($*2.50*)

Subjects and Complements in HPSG Robert Borsley CSLI–87–107 ($*2.50*)

Tools for Morphological Analysis Mary Dalrymple, Ronald M. Kaplan, Lauri Karttunen, Kimmo Koskenniemi, Sami Shaio, Michael Wescoat CSLI–87–108 ($*10.00*)

Cognitive Significance and New Theories of Reference John Perry CSLI–87–109 ($*2.00*)

Fourth Year Report of the Situated Language Research Program CSLI–87–111 (*free*)

Bare Plurals, Naked Relatives, and Their Kin Dietmar Zaefferer CSLI–87–112 ($*2.50*)

Events and "Logical Form" Stephen Neale CSLI–88–113 ($*2.00*)

Backwards Anaphora and Discourse Structure: Some Considerations Peter Sells CSLI–87–114 ($*2.50*)

Toward a Linking Theory of Relation Changing Rules in LFG Lori Levin CSLI–87–115 ($*4.00*)

Fuzzy Logic L. A. Zadeh CSLI–88–116 ($*2.50*)

Dispositional Logic and Commonsense Reasoning L. A. Zadeh CSLI–88–117 ($*2.00*)

Intention and Personal Policies Michael Bratman CSLI–88–118 ($*2.00*)

Propositional Attitudes and Russellian Propositions Robert C. Moore CSLI–88–119 ($*2.50*)

Unification and Agreement Michael Barlow CSLI–88–120 ($*2.50*)

Extended Categorial Grammar Suson Yoo and Kiyong Lee CSLI–88–121 ($*4.00*)

The Situation in Logic—IV: On the Model Theory of Common Knowledge Jon Barwise CSLI–88–122 ($*2.00*)

Unaccusative Verbs in Dutch and the Syntax-Semantics Interface Annie Zaenen CSLI–88–123 ($*3.00*)

What Is Unification? A Categorical View of Substitution, Equation and Solution Joseph A. Goguen CSLI–88–124 ($*3.50*)

Types and Tokens in Linguistics Sylvain Bromberger CSLI–88–125 (*$3.00*)

Determination, Uniformity, and Relevance: Normative Criteria for Generalization and Reasoning by Analogy Todd Davies CSLI–88–126 (*$4.50*)

Modal Subordination and Pronominal Anaphora in Discourse Craige Roberts CSLI–88–127 (*$4.50*)

The Prince and the Phone Booth: Reporting Puzzling Beliefs Mark Crimmins and John Perry CSLI–88–128 (*$3.50*)

Lecture Notes

The titles in this series are distributed by the University of Chicago Press and may be purchased in academic or university bookstores or ordered directly from the distributor at 5801 Ellis Avenue, Chicago, Illinois 60637.

A Manual of Intensional Logic Johan van Benthem, second edition. Lecture Notes No. 1

Emotions and Focus Helen Fay Nissenbaum. Lecture Notes No. 2

Lectures on Contemporary Syntactic Theories Peter Sells. Lecture Notes No. 3

An Introduction to Unification-Based Approaches to Grammar Stuart M. Shieber. Lecture Notes No. 4

The Semantics of Destructive Lisp Ian A. Mason. Lecture Notes No. 5

An Essay on Facts Ken Olson. Lecture Notes No. 6

Logics of Time and Computation Robert Goldblatt. Lecture Notes No. 7

Word Order and Constituent Structure in German Hans Uszkoreit. Lecture Notes No. 8

Color and Color Perception: A Study in Anthropocentric Realism David Russel Hilbert. Lecture Notes No. 9

Prolog and Natural-Language Analysis Fernando C. N. Pereira and Stuart M. Shieber. Lecture Notes No. 10

Working Papers in Grammatical Theory and Discourse Structure: Interactions of Morphology, Syntax, and Discourse M. Iida, S. Wechsler, and D. Zec (Eds.) with an Introduction by Joan Bresnan. Lecture Notes No. 11

Natural Language Processing in the 1980s: A Bibliography Gerald Gazdar, Alex Franz, Karen Osborne, and Roger Evans. Lecture Notes No. 12

Information-Based Syntax and Semantics Carl Pollard and Ivan Sag. Lecture Notes No. 13

Non-Well-Founded Sets Peter Aczel. Lecture Notes No. 14

Partiality, Truth and Persistence Tore Langholm. Lecture Notes No. 15

A Logic of Attribute-Value Structures and the Theory of Grammar Mark Johnson. Lecture Notes No. 16

[illegible]

Partiality, Truth and Persistence is a study in partial model theory [illegible]

[illegible]

Tore Langholm is a research fellow in mathematics at the University of Oslo. He is a co-author of *Situations, Language and Logic.*

CSLI Lecture Notes report new developments in the study of language, information, and computation. In addition to lecture notes, the series includes monographs and conference proceedings. Our aim is to make new results, ideas, and approaches available as quickly as possible.

Langholm / Partiality, Truth, & Persistenc
ISBN 0-937073-34-2
9 780937 073346